Random Forests

RANDOM FORESTS

Yu.L. Pavlov

///VSP///
Utrecht • Boston • Köln • Tokyo
2000

VSP BV
P.O. Box 346
3700 AH Zeist
The Netherlands

Tel: +31 30 692 5790
Fax: +31 30 693 2081
vsppub@compuserve.com
www.vsppub.com

First published in 2000

ISBN 90-6764-314-9

Printed in The Netherlands by Ridderprint bv, Ridderkerk.

CONTENTS

PREFACE

The role of probabilistic methods in discrete mathematics cannot be overestimated. By defining the probability measure on a set of the studied combinatorial objects various numerical characteristics of these objects can be considered as random variables and studied using the methods of probability theory. The advantage of this approach is the well-developed probabilistic analytic techniques that allow us in many cases to obtain results, the proof of which by other methods appears too complicated, if indeed it is at all possible. The application of probabilistic methods is connected with extensive use of the terminology of probability theory. The reader will easily understand however that one speaks in fact about solving enumerative problems of discrete analysis. One of the primary research lines is the study of the limit properties of combinatorial objects manifested at the unlimited increase of the number of elements comprising such objects. It is often possible to represent the distributions of the characteristics of combinatorial objects as conditional distributions of the sums of independent random variables so that they can be studied using asymptotic methods in probability theory, namely limit theorems for sums of independent random variables.

Probability analysis of discrete objects was first carried out by V. L. Goncharov in the papers [24, 25] where permutations with increasing degree were considered (see also [40]). Since then a great number of publications have appeared. In this text we will mention only a few monographs which we believe the most noteworthy, namely [5, 21, 38, 43, 44, 48, 51, 78–80]. One research line that has become established today is the study of random graphs. The Russian mathematical school beginning with V. L. Goncharov and V. E. Stepanov has played an important role in the development of this research [38, 40, 43, 78–80, 83].

This book deals with random forests. Our ambition was to show that, despite the outward simplicity of the forest graph design, the problems emerging in relation to the phenomenon are quite challenging, and their solution often requires subtle mathematical methods. Trees and forests are a convenient means of modelling various objects. Among innumerable possible applications we will only name algorithm analysis [48], methods of applied statistics [19], electricity network modeling [15], and random equation theory [41, 42] as examples. Trees and forests are often subgraphs of other graphs with a more complicated structure. Their study is therefore quite useful for the graph theory in general. A well developed trend in this context is the study of random single-valued mappings of a finite set into itself. These mappings can be represented as a directed graph with one arc leading from each vertex and joining the vertex to its image. Hence, any connected component of a mapping graph contains exactly one cycle, the cyclic vertices being the tree roots. If the arcs joining the cyclic vertices are removed, the resulting subgraph will be a forest of rooted trees. Therefore, results obtained for random forests can be used to study random mappings. A systematic account of random mapping theory is given in V. F. Kolchin's monograph [38]. In [52, 53] the author used limit theorems about the maximum size of a tree in a random forest to study the corresponding characteristics of random mappings. Similar results were obtained in [59] for random mappings with constraints on the number of cycles. Let us mention the fascinating problem of random graph evolution [5, 11, 39, 47, 51, 85]. The essence of the problem is the consideration of a random graph in which the number of vertices and edges tends to infinity. The structure of such graph largely depends on the number of edges over number of vertices ratio. If the ratio tends to zero then, with the probability tending to one, the random graph is a forest. If the ratio increases, the graph first acquires connected components with one cycle, and then the structure abruptly grows more complicated. Thus, results of the random forest and mapping theory can be used in the study of the evolution of random graphs. New trends in the study of random trees and forests have recently been developed; their new types and new numeric characteristics are studied, the set of methods employed is expanded and new applications are discovered for the results obtained (e.g. see [1–3, 23, 34, 35, 45, 46, 71, 72, 82]).

The author is a student of Prof. V. F. Kolchin, whose ideas and methods have much influenced the way in which the material is presented. This volume can be considered as a direct continuation and further development of the investigation into the problems discussed in V. F. Kolchin's renowned monograph *Random Mappings* [38]. It is a revised and amplified variant of the Russian version [66], the synopsis of which can be found in [68].

The main probability methods used in the book are the generalized allocation scheme and methods of the branching process theory. In most cases, random forest and mapping problems are confined to the consideration of conditional distributions for sums of auxiliary independent random variables. To solve these problems one has to find both integral and local convergence of the distributions of such sums to limit laws. The diversity of the cases emerging in this relation due to variations in the behaviour of the parameters necessitates the consideration of array schemes and large deviations. While the respective theory of limit theorems does not cover all encountered variants of the parameter modifications (e.g. see [50]), the book is supplied with the necessary proofs.

The study of random forests of rooted trees with labelled vertices began in [52, 53] where a generalized allocation scheme was used to this end. The application of the methods of branching process theory to the study of random forests was preceded by the consideration of random trees with labelled vertices and the corresponding Galton–Watson processes with Poisson distribution of the number of descendants of each particle. This possibility was first mentioned in the paper by V. E. Stepanov [84]. Systematic studies were undertaken in [36, 37] where random trees with limitations on vertex outdegree were also considered. The obvious connection between Galton–Watson branching processes and the families of trees where the probability of each tree equals the probability of the corresponding realization of the process (see [4]) has demonstrated that possible applications of the theory of branching processes to the study of random trees are not limited to labelled trees. The paper [49] introduces the notion of a simply generated family of trees, which includes trees of different kinds, and gives some probability characteristics of such trees. In [61–63], the relationship between Galton–Watson processes with the geometric distribution of the number of direct offspring of each particle and plane planted trees is revealed, including cases with limitations on the outdegree of vertices. We would also like to point out the work by V. A. Vatutin [87] which investigates the trajectory-wise correspondence between random trees and branching processes.

Branching processes were first used to study random forests of rooted trees with labelled vertices in [55, 57]. In [32], such forests were considered under limitations on the outdegree of vertices. The papers [7–9] dealt with random forests comprised of unrooted trees with labelled vertices.

Results for random forests of plane planted trees were first obtained in [89]. The similarity between limit distributions of the numeric characteristics of trees and forests of different classes, and the methods used to obtain the results triggered the development of general theorems which remain true under minimum constraints on tree and forest classes. Where in the past extensive use of the specific distributions of the particle offspring number of the branching process corresponding to the forest class in question was made to prove the results concerning these characteristics, new developments have enabled us to overcome technical complexities and consider the general situation.

The book deals with forests formed by simply generated trees. While the notion of the simply generated family of trees covers many known tree classes, this approach to the study of random forests gives us a chance to obtain results for various forest types in a uniform way. Forests in this case are however not necessarily equiprobable.

The main task of the book is the full description of the limit behaviour of the random forest's most important characteristics — maximum tree size and number of trees of a given size and height. The text does not cover all known data on random forests. It mainly includes the results obtained by the author himself, whereas the results of other authors are used only when they are necessary to prove the major theorems.

The study of random graphs, the results of which are presented in the book, is one of the research activities of the Institute of Applied Mathematical Researches, Karelian Research Centre, Russian Academy of Sciences. The investigation was made possible by constant help from V. Ya. Kozlov and V. F. Kolchin. The content of the activities was largely decided upon at traditional International Petrozavodsk Conferences 'Probability methods in discrete mathematics' [74, 75]. We would also like to stress the significant role of the journal *Discrete mathematics and applications* which helps mathematicians in Russia collaborate in the study of combinatorial objects.

The book consists of four chapters. The first chapter is an auxiliary one, made up mainly of the graph theory and probability theory details necessary for further narration, and of examples of specific random forests. In chapter two, limit distributions of the maximum tree size are obtained; chapter three deals with the limit theorems for the number of trees of a given size; in chapter four, limit behaviour of the random forest height is studied. References to the literary sources are given in the last sections of each chapter. They also contain results supplementing the main content of the sections.

The following pattern is observed in the numbering of theorems, lemmas, corollaries, and formulae in the book. Each section of each chapter uses the numbering of its own for these objects expressed by one number. References within one section use this numbering only. References to an object from another section within one chapter include the number of the section in front of the main number. If an object from a different chapter is referred to the number of the chapter is added. For example, theorem 5 from section 3 of chapter 1 is referred to as theorem 3.5 in other sections of the same chapter, or as theorem 1.3.5 – in other chapters. Similarly, when referring to a section from another chapter the number of the chapter is put in front of the section number.

Petrozavodsk, 1999 Yu. L. Pavlov

CHAPTER 1

PROBABILISTIC METHODS IN STUDYING TREES AND FORESTS

1. Trees and forests

Tree and forest are two of the simplest and best-known notions of graph theory. Let us introduce some necessary definitions.

A *graph* G with n vertices is a pair $\{V, X\}$ such that V is a finite non-empty set with n elements and X is a set of different non-ordered pairs of different elements of V. Elements of V are called *vertices* and elements of X are called *edges.*

If $x = \{v_1, v_2\} \in X$, where $v_1, v_2 \in V$, then we say that the edge x connects v_1 and v_2. It is obvious that this definition of graph eliminated loops (i. e. edges that connect a vertex with itself) and multiple (parallel) edges. Therefore each edge can be met in X only one time.

If an edge x connects vertices v_1 and v_2, then these vertices are called *adjacents.* The vertex v_1 and the edge x are called *incidented.* It is obvious that v_2 and x are incidented too.

The *subgraph* of a graph G is a graph such that its vertices and edges belong to G.

The *route* in the graph is a sequence $v_1, x_1, v_2, x_2, \dots, v_{n-1}, x_{n-1}, v_n$ where $v_1, \dots,$ $v_n \in V$, $x_1, \dots, x_{n-1} \in X$ and $x_1 = \{v_1, v_2\}, x_2 = \{v_2, v_3\}, \dots,$ $x_{n-1} = \{v_{n-1}, v_n\}$. Such route connects v_1 and v_n. A route is called *a chain* if it contains only different edges. A chain is called a *simple chain* if it contains only different vertices. A chain is said to be a *cycle* if $v_1 = v_n$.

The graph G is called *connected* if any two vertices are connected by a simple chain. It is clear that the set V can be presented as $V = V_1 \cup V_2 \cup \cdots \cup V_k$, $k \geqslant 1$, where $V_i \cap V_j = \emptyset$ if $i \neq j$ and any two vertices are connected if and only if these vertices belong to one of the subsets $V_1, \dots, V_k$. Therefore the set of edges X is a union of non-intersecting sets $X = X_1 \cup \cdots \cup X_k$, where edges from X_i are incident only on vertices from V_i, $i = 1, \dots, k$. This means that the graph G consists of subgraphs $\{V_1, X_1\}, \dots, \{V_k, X_k\}$. Each of these subgraphs is called a *connected component* of G.

The *forest* is a graph without cycles. The *tree* is a connected graph without cycles. It is obvious that the forest is a graph such that its every connected component is a tree.

The *rooted tree* is a tree if it has one distinguished vertex. This vertex is called *its root.* The graph is *directed* if every edge is an ordered pair of vertices. In this case an every edge $x = \{v_1, v_2\}$ is called an *arc* or *directed edge* out of v_1 to v_2.

We will consider forests consisting of N rooted directed trees where arcs are directed from roots. We say that the *outdegree* of a vertex is the number of edges emanating from this vertex. A non-root vertex with the a null outdegree is called the *end vertex* or *leaf.*

The *path* in a directed graph is a sequence $v_1, x_1, \ldots, v_{n-1}, x_{n-1}, v_n$ where arc x_i directs from v_i to v_{i+1}, $i = 1, 2, \ldots, n-1$.

The *height* of a vertex is the number of edges in the path from the root to this vertex. The set of forest vertices with the height t is called *t-th stratum* of the forest.

The *tree height* is the maximum height of tree vertices. The *forest height* is the maximum height of the forest trees.

The *volume* or *size* of a tree is the number of tree vertices.

The rooted tree with n non-root vertices is called *labelled* or a *tree with numbered vertices* if all its non-root vertices have numbers $1, 2, \ldots, n$.

The class of forests considered in the book will be described in Section 4. We will now consider only some examples of such forests.

Example 1. Let $\mathfrak{F}'_{N,n}$ be a set of different forests with N rooted trees and n non-root vertices such that roots have the numbers $1, 2, \ldots, N$ and non-root vertices have the numbers $1, 2, \ldots, n$. An example of $\mathfrak{F}'_{N,n}$ with $N = 3, n = 7$ is given in Fig.1. In this and further figures non-root vertices are marked with circles, and roots with circles with points.

Figure 1. Example of a forest from $\mathfrak{F}'_{3,7}$.

In the forest in Fig.1 the volumes of trees with roots $1, 2, 3$ are equal to $4, 1, 5$ respectively. The non-root vertices $1, 2, 6$ form the first stratum and the second stratum is the set $\{3, 4, 5, 7\}$. The height of the second tree is null and the heights of the first and third trees are two: hence the height of the forest is two.

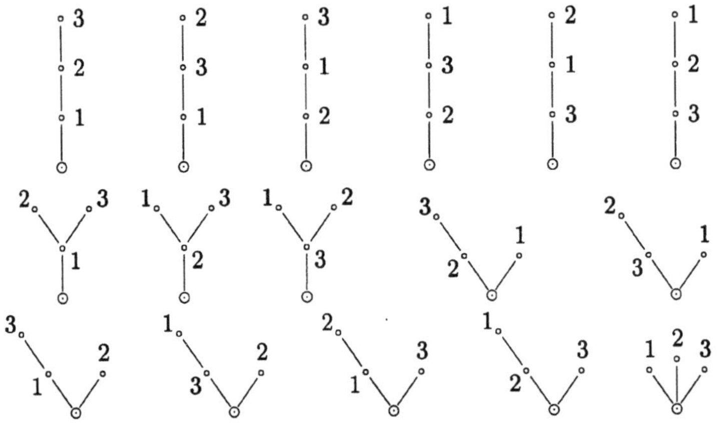

Figure 2. Rooted trees with three labelled non-root vertices.

If $N = 1$ in $\mathfrak{F}'_{N,n}$ then $\mathfrak{F}'_{1,n}$ is a tree. The number of different rooted trees with n non-root labelled vertices is equal to $(n+1)^{n-1}$. This fact is the corollary of the Theorem 1 proved below. All different trees with three labelled non-root vertices are illustrated in Fig.2.

Example 2. Let R be some non-empty set of whole non-negative numbers and $0 \in R$. We denote by $\mathfrak{F}'_{N,n}(R)$ the set of forests in which outdegrees of vertices belong to R. Figure 3 shows an example of the forest with so-called binary trees, where $R = \{0, 2\}$.

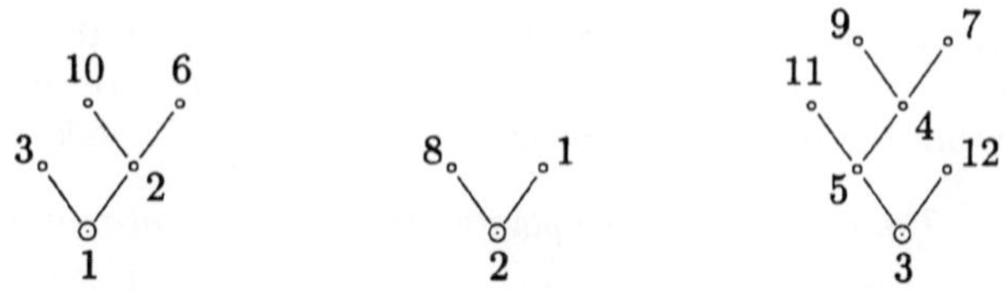

Figure 3. Example of a forest from $\mathfrak{F}'_{3,12}(\{0,2\})$.

Two trees are called *isomorphic* if there exists a biunique correspondence between their sets of vertices with the adjacency preserved.

Non-labelled n-vertex rooted tree is the class of isomorphic rooted trees with n vertices. All non-labelled 4-vertex rooted trees are illustrated in Fig.4.

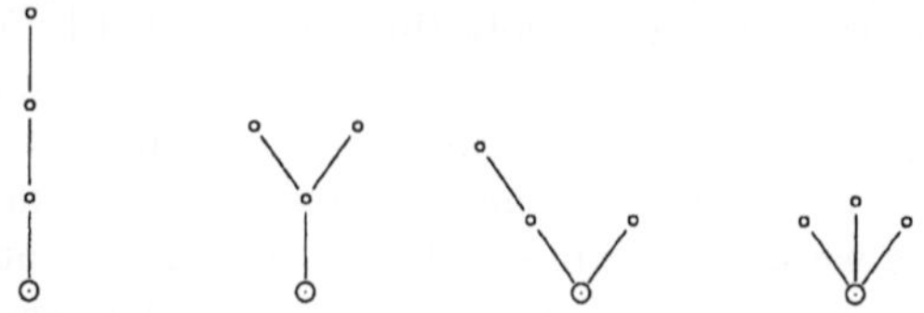

Figure 4. Non-labelled 4-vertex rooted trees.

A rooted tree is called *planted* if the root outdegree is one. All non-labelled 5-vertex planted trees are shown in Fig.5.

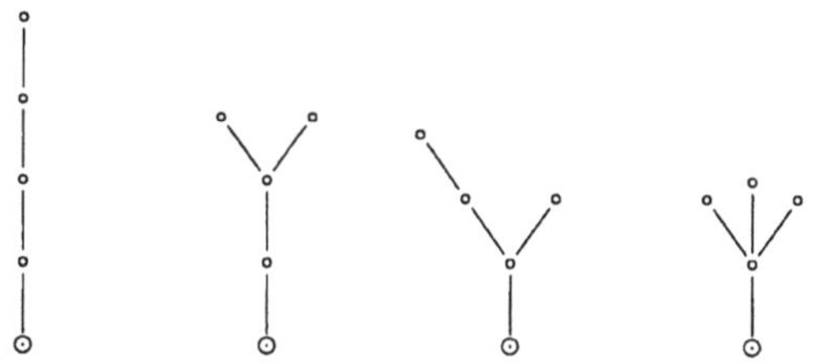

Figure 5. Non-labelled 5-vertex planted trees.

The tree is called *plane* if it is put in the Euclidean plane. It means that in such tree there exists cyclic ordering of the edges incident to every vertex. We are considering trees with arcs directed from roots. Therefore, there exists linear ordering of the arcs incident to any non-root vertex. But in root ordering of arcs is cyclic. To see linear ordering for all vertices we can consider plane planted trees. All different plane planted trees with four non-root non-labelled vertices are illustrated in Fig.6.

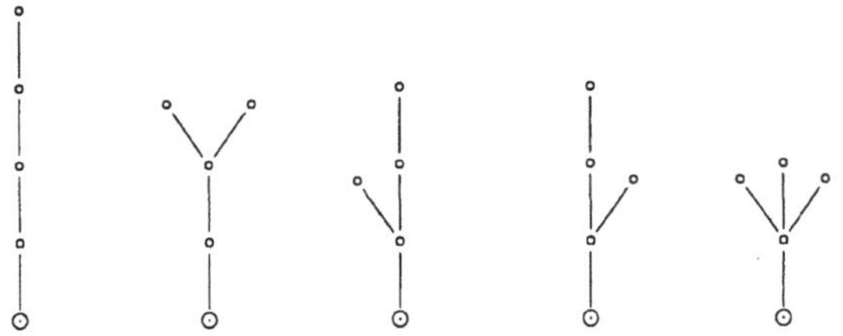

Figure 6. Plane planted 5-vertex trees.

We see that any plane planted tree has a single vertex such that it is incident on the root. This vertex is called the *main node.* To make it more convenient we will assume below that the root is invisible and consider the main node as the root.

Lemma 1. *The number of plane planted trees with n non-root non-labelled vertices is equal to*

$$\frac{1}{n+1}\binom{2n}{n}.$$

Proof. We can compose the number code for any plane planted tree in the following way. Let us move from the root along the first arc to the vertex of the first stratum. We denote this movement $+1$. If the reached vertex is not a leaf we then move up again along the first arc emanting from this vertex to the vertex of the second stratum and again denote this movement $+1$. We continue moving in this way until we arrive at the leaf. After this we move down to the vertex of the preceding stratum. We denote this movement -1. If there are non-passing arcs emanating from this vertex then we move up along the first one of them and denote this $+1$. Otherwise we move down and denote this movement -1. We finish the coding if we attend all vertices and return to the root. Thus every plane planted tree with n non-labelled non-root vertices corresponds to code $\tilde{\alpha}_n = (\alpha_1, \alpha_2, \ldots, \alpha_{2n})$, where $\alpha_1 = +1$, $\alpha_{2n} = -1$, $\alpha_i = \pm 1$, $2 \leqslant i \leqslant 2n-1$. The tree with the code $(1,-1,1,1,1,-1,1,1,-1,-1,-1,-1,1,-1)$ is given in Fig.7.

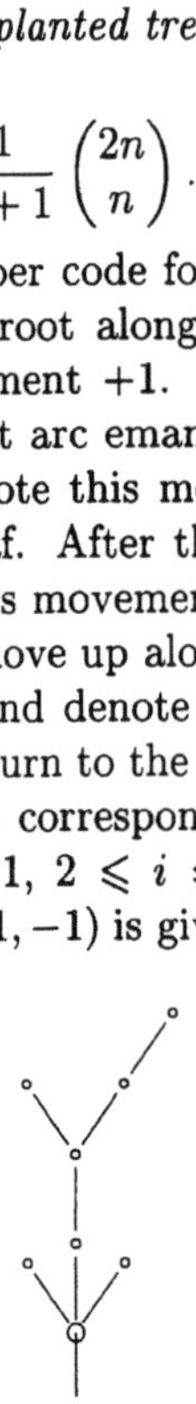

Figure 7. Plane planted 8-vertex tree.

It is readily seen that for every such code there exists a plane planted tree. It follows that there is biunique correspondence between the set of plane planted trees and the set of all codes such as $\tilde{\alpha}_n$. Hence the number of trees is equal to the number of codes $(\alpha_1, \ldots, \alpha_{2n})$, where $\alpha_i = \pm 1$, $1 \leqslant i \leqslant 2n$,

$$\sum_{i=1}^{k} \alpha_i \geqslant 0, \quad k = 1, 2, \ldots, 2n, \quad \sum_{i=1}^{2n} \alpha_i = 0.$$

Let C_n be the number of such codes. Clearly $C_1 = 1$, $C_2 = 2$. Let $\tilde{C}_n$ be the number of codes with

$$\sum_{i=1}^{k} \alpha_i > 0, \quad k = 1, 2, \ldots, 2n-1, \quad \sum_{i=1}^{2n} \alpha_i = 0. \tag{1}$$

Let $\tilde{\alpha}_n$ be such that

$$\sum_{i=1}^{2s-1} \alpha_i > 0, \quad \sum_{i=1}^{2s} \alpha_i = 0, \quad 1 \leqslant s \leqslant n,$$

where s is the minimal number with this property. If $s = n$, then the number of such codes is $\tilde{C}_n$. But if $1 \leqslant s \leqslant n-1$, then this number is $\tilde{C}_s C_{n-s}$. Then

$$C_n = \tilde{C}_n + \sum_{s=1}^{n-1} \tilde{C}_s C_{n-s}. \tag{2}$$

If the code $\tilde{\alpha}_n$ satisfies (1), then taking into consideration that $\alpha_1 = 1$, $\alpha_{2n} = -1$, we find out that for the code $(\alpha_2, \ldots, \alpha_{2n-1})$

$$\sum_{i=2}^{k} \alpha_i \geqslant 0, \quad k = 2, 3, \ldots, 2n-2, \quad \sum_{i=2}^{2n-1} \alpha_i = 0$$

and $\tilde{C}_n = C_{n-1}$, $\tilde{C}_1 = \tilde{C}_0 = 1$. From these relations and (2) we obtain the recurrent relation

$$C_n = \sum_{k=0}^{n-1} C_k C_{n-k-1}, \quad C_0 = 1. \tag{3}$$

We consider the generating function

$$C(z) = \sum_{n=0}^{\infty} C_n z^n. \tag{4}$$

Since $C_n \leqslant 2^{2n}$, the series (4) is converged if $|z| < 1/4$. Multiplying (3) by z^n and summing by n from 1 to ∞ we obtain

$$\sum_{n=1}^{\infty} C_n z^n = \sum_{n=1}^{\infty} z^n \sum_{k=0}^{n-1} C_k C_{n-k-1}.$$

Changing the summation index $s = n - 1$ in the right side of the last equality we obtain

$$zC^2(z) - C(z) + 1 = 0.$$

Solving this square equation we see that, if $|z| < 1/4$ and $C(0) = 1$, then

$$C(z) = (2z)^{-1} \left[1 - (1-4z)^{1/2}\right]. \tag{5}$$

From this relation

$$C(z) = 2^{-1} \sum_{n=0}^{\infty} (-1)^n \binom{1/2}{n+1} 4^{n+1} z^n.$$

Therefore $C_n = (-1)^n \binom{1/2}{n+1} 2^{2n+1}$. After elementary transformations we obtain

$$C_n = \frac{1}{n+1}\binom{2n}{n}.$$

Lemma 1 is proved.

We consider forests which consist of plane planted trees.

Example 3. We denote by $\mathfrak{F}''_{N,n}$ the set of all different forests which consist of N plane planted trees and n non-root vertices where roots have numbers $1, 2, \ldots, N$, but non-root vertices are non-labelled.

Example 4. By analogy with example 2 we can consider the set $\mathfrak{F}''_{N,n}(R)$ which consists of all forests from $\mathfrak{F}''_{N,n}$ such that the outdegrees of vertices belong to R.

Further it will be useful to know the numbers of elements in sets $\mathfrak{F}'_{N,n}$ and $\mathfrak{F}''_{N,n}$.

Theorem 1. *The number of forests from $\mathfrak{F}'_{N,n}$ is equal to $N(N+n)^{n-1}$.*

Proof. Let us choose any forest from $\mathfrak{F}'_{N,n}$. For this forest we consider the next allocation of $n-1$ labelled particles to $N+n$ cells, where N cells have numbers $1, 2, \ldots, N$ and correspond to N roots of the forest. Other n cells will be considered separately. They have the numbers $1', 2', \ldots, n'$. Let us find the leaf with the maximum number in the forest and let k_1 be the number of the adjacent vertex. Then put the particle with the number 1 in the cell with the number k'_1. If k_1 is the number of the root we put the first particle in the cell with the number k_1. Thereafter we remove from the forest the leaf with the maximum number and the arc incident to this leaf. Further again we find the leaf with the maximum number and let k_2 be the number of the adjacent vertex. We put the particle with the number 2 in the cell with the number k'_2 (or k_2) and again remove the leaf and the incident arc. We will continue this process to the last arc of the forest. Finally we obtain the concrete allocation of $n-1$ particles to $N+n$ cells. Note that in this allocation there remain empty cells. These cells correspond to roots without arcs and a single root with the last arc. Also empty cells correspond to leaves.

On the contrary let us have any allocation of $n-1$ labelled particles to $N+n$ cells. For this allocation we can find a corresponding forest. Let us find the empty cell with the maximum number m'_1 among the cells with numbers $1', 2', \ldots, n'$. Let m_1 be the number of cell with the particle 1. Then we consider the vertex with the number m'_1 as the leaf with the adjacent vertex m_1. After this we remove the cell m_1 and the particle 1. In the same way we can find the empty cell with the maximum number m'_2 and join the vertex m_2 with the vertex m'_2 by an arc, where m_2 is the number of the cell with the particle 2. We will continue this process until we remove all $n-1$ particles. After this we will have N empty root cells and one of the n non-root cells with the number of the last vertex. We can join this vertex to one of N roots. It means that for this specific allocation we can construct N different forests from $\mathfrak{F}'_{N,n}$. It is well-known that there exists $(N+n)^{n-1}$ different allocations of $n-1$ particles to $N+n$ cells. Thus there are $N(N+n)^{n-1}$ different forests.

Theorem 2. *The number of forests from $\mathfrak{F}''_{N,n}$ is equal to*

$$\frac{N}{N+n}\binom{2n+N-1}{n}.$$

Proof. We denote by $L(N,n)$ the number of different forests from $\mathfrak{F}''_{N,n}$. Since $L(1,n)$ is the number of different $(n+1)$-vertex plane planted trees the assertion of Theorem 2 for $N=1$ is the corollary of Lemma 1. Let us show that the equality

$$L(2,n) = L(1,n+1) \tag{6}$$

is valid. For this we represent $L(2,n)$ as

$$L(2,n) = \sum_{k=0}^{n} L(1,k)L(1,n-k).$$

From this relation and Lemma 1 it follows that

$$L(2,n) = \sum_{k=0}^{n} \left(\frac{1}{k+1}\binom{2k}{k}\right)\left(\frac{1}{n-k+1}\binom{2n-2k}{n-k}\right). \tag{7}$$

Taking in (3) $n+1$ instead of n from Lemma 1 we obtain

$$\frac{1}{n+2}\binom{2n+2}{n+1} = \sum_{k=0}^{n} \left(\frac{1}{k+1}\binom{2k}{k}\right)\left(\frac{1}{n-k+1}\binom{2n-2k}{n-k}\right).$$

From this and (7) it follows that

$$L(2,n) = \frac{1}{n+2}\binom{2n+2}{n+1} = L(1,n+1).$$

This relation implies (6). Using (6) it is easy to prove the recurrent relation

$$L(N,n) = L(N-1,n+1) - L(N-2,n+1). \tag{8}$$

Really

$$L(N,n) = \sum_{k=0}^{n} L(2,k)L(N-2,n-k)$$

$$= \sum_{k=0}^{n} L(1,k+1)L(N-2,n-k) = L(N-1,n+1)$$

$$-L(1,0)L(N-2,n+1) = L(N-1,n+1) - L(N-2,n+1).$$

The assertion of Theorem 2 follows from Lemma 1, (6) and (8) by induction.

2. Random forests and generalized allocation scheme

The main idea of the book is the use of probabilistic methods in studying forests. We will shortly remind the reader of the main probability notions.

Let Ω be a set of arbitrary elements. If $\omega \in \Omega$ then we say that ω is the *elementary event*. The set Ω is called the *space of elementary events*. Let $\mathfrak{A}$ be the σ-algebra over Ω. This means that $\mathfrak{A}$ is the set of subsets Ω such that: 1) $\Omega \in \mathfrak{A}$; 2) if $A_i \in \mathfrak{A}$, $i = 1, 2, \ldots$, then $\bigcup_{i=1}^{\infty} A_i \in \mathfrak{A}$; 3) if $A \in \mathfrak{A}$ then $\Omega \setminus A \in \Omega$. Let also $\mathbf{P}$ be a non-negative countably additive function on $\mathfrak{A}$ such that $\mathbf{P}\{\Omega\} = 1$. This function is called *probability*. The three $(\Omega, \mathfrak{A}, \mathbf{P})$ is called *probability space*.

A *random variable* is the real measurable function $\xi = \xi(\omega)$, $\omega \in \Omega$. The *distribution function* of the random variable ξ is function $F_\xi(x)$ such that $F_\xi(x) = \mathbf{P}\{\xi < x\}$.

Let $\varphi(\omega)$ be the map from Ω to Y, where Y is an arbitrary set. This function, which is a generalization of a random variable, is called the ***random element*** of Y.

Now probabilistic methods for studying combinatorial problems are sufficiently well developed. In such problems, Ω usually denotes a set of combinatorial objects. We determine the probability $\mathbf{P}\{\omega\}$ for each $\omega \in \Omega$. The set Ω is finite; therefore the probability exists for every subset of Ω. It is clear that any number characteristic of ω is a random variable. We will consider the random element of Ω as the identical function $\varphi(\omega) = \omega$, $\omega \in \Omega$. We can interpret such random element as the random choice of the element $\omega \in \Omega$ corresponding to the probability $\mathbf{P}$.

Let $\Omega = \mathfrak{F}$ be some set of forests. ***Random forest*** F is the identical map of the set $\mathfrak{F}$ into itself such that $\mathbf{P}\{F = f\} = \mathbf{P}\{f\}$ for any $f \in \mathfrak{F}$. It follows that a random forest is the random element of $\mathfrak{F}$.

For example, if we consider uniform distribution on the set of forests, then from Theorems 1.1 and 1.2 we obtain

$$\mathbf{P}\{F = f\} = (N(N+n)^{n-1})^{-1}, \quad f \in \mathfrak{F}'_{N,n};$$
$$\mathbf{P}\{F = f\} = (N+n)/\left(N\binom{2n+N-1}{n}\right), \quad f \in \mathfrak{F}''_{N,n}.$$

A ***random tree*** is a random forest under the condition $N = 1$.

We will use mainly two probabilistic methods to study random forests. The first one is the generalized allocation scheme and the second one is the use of the branching process theory. We will now consider some applications of the first method. The examples of the second method will be considered in Section 3.

The generalized allocation scheme is given as follows. Let $\xi_1, \dots, \xi_N$ be independent identically distributed random variables. Let also $\eta_1, \dots, \eta_N$ be random variables such that $\eta_1 + \dots + \eta_N = n$ and

$$\begin{aligned}&\mathbf{P}\{\eta_1 = k_1, \dots, \eta_N = k_N\}\\ &= \mathbf{P}\{\xi_1 = k_1, \dots, \xi_N = k_N \mid \xi_1 + \dots + \xi_N = n\}.\end{aligned} \tag{1}$$

The name of this method can be explained in the following way. The relation (1) can be interpreted as the allocation of n particles to N cells. Then the random variable η_i is the number of particles in the cell number i after the allocation, $1 \leqslant i \leqslant N$. The simplest example of it is the classical allocation scheme with equiprobable allocation of particles. In this case if $k_1 + \dots + k_N = n$ then

$$\mathbf{P}\{\eta = k_1, \dots, \eta_N = k_N\} = \frac{n!}{k_1! \dots k_N! N^n}$$

and relation (1) is valid if

$$\mathbf{P}\{\xi_i = k\} = \frac{\lambda^k}{k!}e^{-\lambda}, \quad k = 0, 1, \dots; \quad i = 1, \dots, N; \quad \lambda > 0.$$

The generalized allocation scheme arises when the allocation of particles is not equiprobable and the distribution of $\xi_1, \dots, \xi_N$ is not Poisson. Let

$$p_k = \mathbf{P}\{\xi_1 = k\}, \quad k = 0, 1, \dots \tag{2}$$

Below we will prove Lemmas 1 and 2 about important characteristics of the generalized allocation scheme. These Lemmas are examples of using the relation (1).

Let $\mu_r(N,n)$ be the cell number with r particles in the scheme (1), (2). We denote by $\xi_1^{(r)},\dots,\xi_N^{(r)}$ auxiliary independent identically distributed random variables such that

$$\mathbf{P}\{\xi_1^{(r)}=k\}=\mathbf{P}\{\xi_1=k|\,\xi_1\neq r\},\quad k=0,1,\dots$$

Let also

$$\zeta_N=\xi_1+\dots+\xi_N,\quad \zeta_N^{(r)}=\xi_1^{(r)}+\dots+\xi_N^{(r)}.$$

Lemma 1. *The equality*

$$\mathbf{P}\{\mu_r(N,n)=k\}=\binom{N}{k}p_r^k(1-p_r)^{N-k}\frac{\mathbf{P}\{\zeta_{N-k}^{(r)}=n-kr\}}{\mathbf{P}\{\zeta_N=n\}}$$

holds.

Proof. From (1) it follows that

$$\mathbf{P}\{\mu_r(N,n)=k\}=\binom{N}{k}\mathbf{P}\{\eta_1\neq r,\dots,\eta_{N-k}\neq r,$$

$$\eta_{N-k+1}=r,\dots,\eta_N=r\}=\binom{N}{k}p_r^k(1-p_r)^{N-k}$$

$$\times\mathbf{P}\{\zeta_N=n|\,\xi_1\neq r,\dots,\xi_{N-k}\neq r,\xi_{N-k+1}=r,\dots,\xi_N=r\}/\mathbf{P}\{\zeta_N=n\}.$$

This relation implies Lemma 1.

We denote by $\eta_{(N)}$ the maximum number of particles in one cell; therefore

$$\eta_{(N)}=\max_{1\leqslant i\leqslant N}\eta_i.$$

Let $\tilde\xi_1^{(r)},\dots,\tilde\xi_N^{(r)}$ be auxiliary independent identically distributed random variables such that

$$\mathbf{P}\{\tilde\xi_1^{(r)}=k\}=\mathbf{P}\{\xi_1=k|\,\xi_1\leqslant r\},\quad k=0,1,\dots$$

and let $\tilde\zeta_N^{(r)}=\tilde\xi_1^{(r)}+\dots+\tilde\xi_N^{(r)}$, $P_r=\mathbf{P}\{\xi_1>r\}$.

Lemma 2. *The equality*

$$\mathbf{P}\{\eta_{(N)}\leqslant r\}=(1-P_r)^N\frac{\mathbf{P}\{\tilde\zeta_N^{(r)}=n\}}{\mathbf{P}\{\zeta_N=n\}}$$

holds.

Proof. Using (1) we obtain

$$\begin{aligned}\mathbf{P}\{\eta_{(N)}\leqslant r\}&=\mathbf{P}\{\eta_1\leqslant r,\dots,\eta_N\leqslant r\}\\&=\mathbf{P}\{\xi_1\leqslant r,\dots,\xi_N\leqslant r|\,\xi_1+\dots+\xi_N=n\}\\&=(\mathbf{P}\{\xi_1\leqslant r\})^N\,\mathbf{P}\{\zeta_N=n|\,\xi_1\leqslant r,\dots,\xi_N\leqslant r\}/\mathbf{P}\{\zeta_N=n\}.\end{aligned}$$

It is readily seen that the Lemma is proved.

Lemmas 1 and 2 show that to obtain the limit distributions of $\mu_r(N,n)$ and $\eta_{(N)}$ it suffices to consider the asymptotic behaviour of

$$\binom{N}{k}p_r^k(1-p_r)^{N-k},\quad (1-P_r)^N$$

and the probabilities $\mathbf{P}\{\zeta_{N-k}^{(r)}=n-kr\}$, $\mathbf{P}\{\zeta_N=n\}$, $\mathbf{P}\{\tilde\zeta_N^{(r)}=n\}$.

We consider the applications of the generalized allocation scheme to the study of $\mathfrak{F}'_{N,n}$ and $\mathfrak{F}''_{N,n}$ (see Examples 1.1 and 1.3) with uniform distributions of probabilities.

Example 1. Using Theorem 1.1 for $N = 1$ we find that the number of forests in $\mathfrak{F}'_{N,n}$ with k_i non-root vertices in the i-th tree, $i = 1, \dots, N$, is equal to

$$\frac{n!}{k_1! \dots k_N!}(k_1+1)^{k_1-1} \dots (k_N+1)^{k_N-1},$$

where $k_1 + \dots + k_N = n$ and $n!/(k_1! \dots k_N!)$ is the number of allocating ways of n non-root vertices to N trees with $k_1, \dots, k_N$ non-root vertices in trees with roots $1, \dots, N$. Applying Theorem 1.1 once more we obtain

$$\mathbf{P}\{\eta'_1 = k_1, \dots, \eta'_N = k_N\} = \frac{n!(k_1+1)^{k_1} \dots (k_N+1)^{k_N}}{N(N+n)^{n-1}(k_1+1)! \dots (k_N+1)!}, \tag{3}$$

where $\eta'_1, \dots, \eta'_N$ are the numbers of non-root vertices in trees of the forest from $\mathfrak{F}'_{N,n}$ with roots $1, \dots, N$ respectively.

We consider independent identically distributed random variables $\xi'_1, \dots, \xi'_N$ such that

$$\mathbf{P}\{\xi'_1 = k\} = \frac{(k+1)^k}{(k+1)!}\frac{x^{k+1}}{\theta(x)}, \quad k = 0, 1, \dots, \tag{4}$$

where x is the number parameter, $0 < x \leqslant e^{-1}$, and the function $\theta(x)$ is

$$\theta(x) = \sum_{k=1}^{\infty} k^{k-1}x^k/k!$$

From this relation we obtain

$$\mathbf{P}\{\xi'_1 + \dots + \xi'_N = n\}$$
$$= \frac{x^{N+n}}{n!\theta^N(x)} \sum_{k_1+\dots+k_N=n} n! \frac{(k_1+1)^{k_1} \dots (k_N+1)^{k_N}}{(k_1+1)! \dots (k_N+1)!}.$$

The last sum is equal to the number of forests in $\mathfrak{F}'_{N,n}$. Therefore from Theorem 1.1 we obtain

$$\mathbf{P}\{\xi'_1 + \dots + \xi'_N = n\} = \frac{N(N+n)^{n-1}}{n!}\frac{x^{N+n}}{\theta^N(x)}. \tag{5}$$

Using this relation and (4) we easily find that

$$\mathbf{P}\{\xi'_1 = k_1, \dots, \xi'_N = k_N | \ \xi'_1 + \dots + \xi'_N = n\}$$
$$= \frac{n!(k_1+1)^{k_1} \dots (k_N+1)^{k_N}}{N(N+n)^{n-1}(k_1+1)! \dots (k_N+1)!}.$$

Comparing the last relation with (3) and (1) we see that for random variables $\eta'_1, \dots, \eta'_N$ and $\xi'_1, \dots, \xi'_N$ the generalized allocation scheme is true. It means that changing (2) to (4) we can use Lemmas 1 and 2 to study the behaviour of the number of trees with r non-root vertices and the maximum tree volume in $\mathfrak{F}'_{N,n}$.

Example 2. According to Lemma 1.1 the number of forests in $\mathfrak{F}''_{N,n}$ with k_i non-root vertices in the i-th tree is

$$\binom{2k_1}{k_1} \dots \binom{2k_N}{k_N} \frac{n!}{(k_1+1)! \dots (k_N+1)!}.$$

Hence by virtue of Theorem 1.2

$$\mathbf{P}\{\eta_1'' = k_1, \dots, \eta_N'' = k_N\}$$

$$= \frac{(N+n)n! \binom{2k_1}{k_1} \cdots \binom{2k_N}{k_N}}{N \binom{2n+N-1}{n} (k_1+1)! \dots (k_N+1)!}, \tag{6}$$

where η_i'' is the number of non-root vertices in i-th tree of a forest from $\mathfrak{F}_{N,n}''$.

From (1.4), (1.5) and Lemma 1.1 it follows that for $|x| < 1/4$

$$\sum_{k=0}^{\infty} \binom{2k}{k} x^k/(k+1) = (1 - \sqrt{1-4x})/(2x). \tag{7}$$

Taking into consideration (7) we denote by $\xi_1'', \dots, \xi_N''$ the independent identically distributed random variables such that

$$\mathbf{P}\{\xi_1'' = k\} = 2 \binom{2k}{k} \frac{x^{k+1}}{(k+1)(1-\sqrt{1-4x})}, \quad k = 0, 1, \dots, \tag{8}$$

where x is the number parameter, $0 < x < 1/4$. From this by analogy with (5) we obtain

$$\mathbf{P}\{\xi_1'' + \dots + \xi_N'' = n\}$$

$$= \frac{2^N x^{N+n}}{n!(1-\sqrt{1-4x})^N} \sum_{k_1+\dots+k_N=n} \binom{2k_1}{k_1} \cdots \binom{2k_N}{k_N} \frac{n!}{(k_1+1)! \dots (k_N+1)!}$$

$$= \binom{2n+N-1}{n} \frac{2^N N x^{N+n}}{n!(N+n)(1-\sqrt{1-4x})^N}.$$

From the last relation and (8) we obtain

$$\mathbf{P}\{\xi_1'' = k_1, \dots, \xi_N'' = k_N | \, \xi_1'' + \dots + \xi_N'' = n\}$$

$$= \frac{(N+n)n! \binom{2k_1}{k_1} \cdots \binom{2k_N}{k_N}}{N \binom{2n+N-1}{n} (k_1+1)! \dots (k_N+1)!}.$$

From this and (6) it follows that as in example 1 we can use Lemmas 1 and 2 to study tree volumes in $\mathfrak{F}_{N,n}''$ if we consider (8) instead of (2).

The reader will easily understand that by analogy we can use the generalized allocation scheme to study forests from $\mathfrak{F}_{N,n}'(R)$, $\mathfrak{F}_{N,n}''(R)$ (see Examples 1.2 and 1.4).

3. Random forests and branching processes

The idea of using branching process theory to study random forests is based on the intuitive image of a tree as a realization of the branching process. The Galton–Watson branching process is most convenient for the formalization of this idea.

Let $\xi_1, \xi_2, \dots$ be independent identically distributed random variables such that

$$\mathbf{P}\{\xi_1 = k\} = p_k, \quad k = 0, 1, \dots \tag{1}$$

The Galton–Watson branching process G starting with N particles is the sequence of random variables $\mu(t)$, $t = 0, 1, \ldots$, if

$$\mu(0) = N, \quad \mu(t+1) = \xi_1 + \cdots + \xi_{\mu(t)} \tag{2}$$

and $\mu(t) = 0$ implies $\mu(t+1) = 0$. We will interpret the random variable $\mu(t)$ as the number of particles in the t-th generation of G. Any particle of the t-th generation in the next moment of time $t+1$ dies and independently of other particles gives birth to k particles of $(t+1)$-th generation, $k = 0, 1, \ldots$ The number of offspring of a particle is a random variable with the distribution (1). That is why this distribution is called *offspring distribution* with the generating function

$$F(z) = \sum_{k=0}^{\infty} p_k z^k. \tag{3}$$

We denote by $F_t(z)$ the generating function of $\mu(t)$. Therefore

$$F_t(z) = \mathbf{E}\, z^{\mu(t)}, \quad t = 0, 1, \ldots \tag{4}$$

Let m be the mathematical expectation of the distribution (1). Then from (2) we obtain

$$\mathbf{E}\,\mu(t+1) = m\,\mathbf{E}\,\mu(t), \quad t = 1, 2, \ldots \tag{5}$$

Let $N = 1$. Using (5) and the equality $\mu(1) = m$ we can prove by induction

$$\mathbf{E}\,\mu(t) = m^t, \quad t = 0, 1, \ldots$$

A branching process starting with one particle is called *subcritical* if $m < 1$, *critical* if $m = 1$ and *supercritical* if $m > 1$. We say that the Galton–Watson process is *extinguised in the moment* t if $\mu(t-1) > 0$, $\mu(t) = 0$. Let

$$q = \mathbf{P}\left\{ \bigcup_{t=0}^{\infty} \{\mu(t) = 0\} \right\}. \tag{6}$$

Futher we will not consider branching processes such that $p_0 + p_1 = 1$ or $p_0 = 0$.

Theorem 1. *Let* $N = 1$. *If* $m \leqslant 1$ *then* $q = 1$, *if* $m > 1$ *then* $q < 1$.

Proof. Using (2) and (4) we obtain

$$F_{t+1}(z) = \sum_{k=0}^{\infty} \mathbf{P}\{\mu(t+1) = k\} z^k = \sum_{k=0}^{\infty} z^k \sum_{i=0}^{\infty} \mathbf{P}\{\mu(t) = i\}$$

$$\times\, \mathbf{P}\{\xi_1 + \cdots + \xi_i = k\} = \sum_{i=0}^{\infty} \mathbf{P}\{\mu(t) = i\} \sum_{k=0}^{\infty} z^k\, \mathbf{P}\{\xi_1 + \cdots + \xi_i = k\}$$

$$= \sum_{i=0}^{\infty} \mathbf{P}\{\mu(t) = i\} (F(z))^i = F_t(F(z)).$$

From this relation and equalities $F_0(z) = z$, $F_1(z) = F(z) = F(F_0(z))$ the reader will have no difficulty in seeing that

$$F_{t+1}(z) = F(F_t(z)), \quad t = 0, 1, \ldots \tag{7}$$

Putting $z = 0$ in this equation we obtain

$$\mathbf{P}\{\mu(t+1) = 0\} = F(\mathbf{P}\{\mu(t) = 0\}). \tag{8}$$

Since $\{\mu(t) = 0\} \subseteq \{\mu(t+1) = 0\}$ it follows from (6) that

$$q = \lim_{t\to\infty} \mathbf{P}\{\mu(t) = 0\}. \tag{9}$$

Hence using (8) as $t \to \infty$ we obtain that q is the solution of the equation

$$x = F(x). \tag{10}$$

The equality $\mathbf{E}\,\mu(1) = m$ shows that $F'(x)$ exists in the interval $[0,1]$ and $F'(1) = m$. It can easily be seen that if $m \leqslant 1$ then $F(0) = p_0$, $F(1) = 1$, $F''(x) > 0$. Therefore $F(x) > x$ in the interval $[0,1)$ and (10) has the only solution $q = 1$. It is clear that as $m > 1$ the equation (10) has two solutions $x_1 < 1$ and $x_2 = 1$. Using obvious relations $\mathbf{P}\{\mu(t) = 0\} = p_0 \leqslant F(x_1) = x_1$ from (8) we obtain by induction $\mathbf{P}\{\mu(t+1) = 0\} = F(\mathbf{P}\{\mu(t) = 0\}) \leqslant F(x_1) = x_1$. It follows that $\mathbf{P}\{\mu(t) = 0\} \leqslant x_1$ for any t and (9) implies $q < 1$. The Theorem is proved.

A branching process is called *extinguished* if $q = 1$. Further we will consider only subcritical and critical branching processes. Thus all processes in this book are extinction processes.

To prove some results concerning random forests we consider asymptotics of the process continuation probability $\mathbf{P}\{\mu(t) > 0\}$ as $t \to \infty$ and $N = 1$. Let $A(z)$ be any probability generating function:

$$A(z) = \sum_{k=0}^{\infty} a_k z^k, \quad A(1) = 1.$$

Lemma 1. *Let $\delta \in (0,1)$. The series*

$$\sum_{k=0}^{\infty} [1 - A(1 - \delta^k)] \tag{11}$$

is converged if and only if the series

$$\sum_{k=1}^{\infty} a_k \ln k \tag{12}$$

is converged.

Proof. It can be easily seen that

$$\sum_{k=2}^{\infty} [1 - A(1-\delta^k)] \leqslant \sum_{k=1}^{\infty} \int_k^{k+1} [1 - A(1 - \delta^x)]dx \leqslant \sum_{k=1}^{\infty} [1 - A(1 - \delta^k)];$$

therefore

$$0 \leqslant \sum_{k=1}^{\infty} [1 - A(1 - \delta^k)] - \int_1^{\infty} [1 - A(1 - \delta^x)]dx \leqslant 1 - A(1 - \delta).$$

From this it follows that the series (11) converges or diverges simultaneously with the integral

$$\int_1^\infty [1 - A(1 - \delta^x)]dx = -\frac{1}{\ln \delta} \int_{1-\delta}^1 \frac{1 - A(y)}{1 - y} dy. \tag{13}$$

The ratio $(1 - A(y))/(1 - y)$ is the generating function of sums $a_i + a_{i+1} + \ldots, i \geqslant 1$. Using this fact and integrating (13) term by term we obtain that this integral and the series

$$\sum_{k=0}^\infty \frac{1}{k+1} \sum_{i=k+1}^\infty a_i$$

converge or diverge simultaneously. The last expression can be transformed into the series

$$\sum_{k=1}^\infty a_k \sum_{i=1}^k i^{-1}$$

which converges or diverges simultaneously with (12) because

$$\sum_{i=1}^k i^{-1} = \ln k + O(1).$$

This completes the proof of Lemma 1.

By $Q(t)$ denote the probability $\mathbf{P}\{\mu(t) > 0\}$. We consider the limit behaviour $Q(t)$ as $t \to \infty$. For positive integer r, let the expression $F^{*r}(z)$ be the r-fold iteration of the function $F(z)$. Hence $F^{*1}(z) = F(z)$, $F^{*2}(z) = F(F(z))$, $F^{*3}(z) = F(F(F(z)))$, etc.

Theorem 2. *Let $N = 1$, $0 < m \leqslant c < 1$ and $t \to \infty$. The equality $Q(t) = Km^t(1 + o(1))$, $0 < K < \infty$ holds if and only if*

$$\mathbf{E}\,\mu(1) \ln \mu(1) < \infty \tag{14}$$

$$K = \lim_{r \to \infty} \frac{1 - F^{*r}(z)}{m^r}. \tag{15}$$

Proof. Putting $z = 0$ to (7) and using the equality $Q(t) = 1 - F_t(0)$ we obtain

$$Q(t + 1) = 1 - F(F_t(0)). \tag{16}$$

It can easily be seen that the function

$$B(z) = (1 - F(z))/(m(1 - z))$$

is a probability generating function. Let $K_t = Q(t)/m^t$. Hence

$$K_{t+1} = B(F_t(0))K_t$$

and

$$K_t = \prod_{i=0}^{t-1} B(F_i(0)).$$

This implies that

$$K = \lim_{t\to\infty} K_t = \prod_{i=0}^{\infty} B(F_i(0)). \tag{17}$$

This infinite product converges if and only if the series

$$\sum_{i=0}^{\infty}[1 - B(F_i(0))] \tag{18}$$

converges. By Theorem of mean as $0 \leqslant z \leqslant 1$

$$1 - F_2(z) = 1 - F(F(z)) = F'(\delta(z))(1 - F(z)), \tag{19}$$

where $F(0) \leqslant F(z) \leqslant \delta(z) < 1$. Since $0 < \beta = F'(F(0)) \leqslant F'(\delta(z)) < m$ it is not hard to get from (19) that

$$\beta(1 - F(z)) \leqslant 1 - F_2(z) \leqslant m(1 - F(z)).$$

By replacing z with $F(z)$ in the last expression and doing as before we obtain that for $t = 3, 4, \ldots$

$$\beta^{t-1}(1 - F(z)) \leqslant 1 - F_t(z) \leqslant m^{t-1}(1 - F(z)).$$

This together with the relation $Q(1) = 1 - F(0)$ implies that

$$B(1 - Q(1)m^{i-1}) \leqslant B(F_i(0)) \leqslant B(1 - Q(1)\beta^{i-1}). \tag{20}$$

By Lemma 1 the series (11) is converged or diverged for any δ, $0 < \delta < 1$. It can easily be seen from (20) that the series (18) converges or diverges simultaneously with the series

$$\sum_{i=0}^{\infty}[1 - B(1 - \delta^i)]$$

for some $\delta \in (0, 1)$. Using Lemma 1 and the evident expression of $B(z)$ we obtain that the series (18) is converged if and only if the sum

$$\sum_{k=1}^{\infty} \ln k \sum_{i=k+1}^{\infty} p_i$$

is converged. This condition is equivalent to (14). It is clear that (17) implies (15). Theorem 2 is proved.

Corollary 1. *Let $F''(1) < \infty$, $t \to \infty$, $m \to 0$. Then*

$$Q(t) = m^t(1 + o(1)).$$

Proof. Since $F''(1) < \infty$, it follows that condition (14) holds. Therefore the assertion of Theorem 2 is valid. To conclude the proof, it remains to prove that $K = 1$. Using the Taylor formula for $F(z)$ we obtain

$$F(z) = 1 + m(z - 1) + R(z)(z - 1)^2,$$

where $0 \leqslant z \leqslant 1$, $0 \leqslant R(z) \leqslant F''(1)/2$. If we replace z by $1 - Q(t)$ we obtain

$$F(1 - Q(t)) = 1 - mQ(t) + R(1 - Q(t))Q^2(t). \tag{21}$$

From (16) and the equality $Q(t) = 1 - F_t(0)$ it follows that

$$Q(t + 1) = 1 - F(1 - Q(t)). \tag{22}$$

By Theorem of mean as $0 \leqslant z \leqslant 1$

$$Q(t+1) = F'(\delta(z))Q(t),$$

where $0 \leqslant 1 - Q(t) < \delta(z) < 1$. Since $F'(\delta(z)) < m$ it follows that

$$Q(t+1) \leqslant mQ(t) \leqslant \ldots \leqslant m^{t+1}. \tag{23}$$

Hence by (21) and (22)

$$Q(t) = 1 - F(1 - Q(t-1))$$
$$= mQ(t-1) - R(1 - Q(t-1))Q^2(t-1) \geqslant mQ(t-1) - F''(1)m^{2t}/2.$$

Estimating $Q(t-1), Q(t-2), \ldots$ by analogy we obtain

$$Q(t) \geqslant m^t - 2^{-1}F''(1)(m^{t+1} + m^{t+2} + \cdots + m^{2t}) = m^t(1 + O(m)).$$

From this relation and (23) we get that $Q(t)/m^t \to 1$ as $t \to \infty$ and $m \to 0$, the Corollary 1 is thus proved.

Theorem 3. *Let* $N = 1$, $m = 1$, $\mathbf{D}\mu(1) = B$. *Then as* $t \to \infty$

$$Q(t) = \frac{2}{Bt}(1 + o(1)).$$

Proof. Let $F''(1)$ be the second left derivative of $F(z)$ as $z = 1$. Using the Taylor formula and (8) we obtain

$$\mathbf{P}\{\mu(t+1) = 0\} = F(1) + F'(1)(\mathbf{P}\{\mu(t) = 0\} - 1)$$
$$+2^{-1}F''(\delta\,\mathbf{P}\{\mu(t) = 0\} + 1 - \delta)(\mathbf{P}\{\mu(t) = 0\} - 1)^2,$$

where $0 \leqslant \delta \leqslant 1$. From this relation we get

$$Q(t+1) = Q(t) - 2^{-1}F''(1 - \delta Q(t))Q^2(t). \tag{24}$$

Since the process G is critical it follows from Theorem 1 that $Q(t) \to$) as $t \to \infty$. Therefore $F''(1 - Q(t)) \to F''(1) = B$ and from (24) we obtain

$$Q(t+1)/Q(t) = 1 - F''(1 - \delta Q(t))Q(t)/2 \to 1. \tag{25}$$

From this relation and (24) we find

$$Q(t+1) = Q(t) - BQ(t)Q(t+1)/2 + \varepsilon(t),$$

where $\varepsilon(t) = [BQ(t)Q(t+1) - F''(1 - \delta Q(t))Q^2(t)]/2$. Dividing both sides by $Q(t)Q(t+1)$ we obtain

$$\frac{1}{Q(t)} = \frac{1}{Q(t+1)} - \frac{B}{2} - \delta(t), \tag{26}$$

where $\delta(t) = -\varepsilon(t)/(Q(t)Q(t+1))$. It is not hard to get from (25) that $\delta(t) \to 0$ as $t \to \infty$. We rearrange (26) as

$$\frac{1}{Q(t+1)} = \frac{1}{Q(t)} + \frac{B}{2} + \delta(t)$$

and sum both sides of this relation over t from 0 to $s-1$. Since $Q(0) = 1 - \mathbf{P}\{\mu(0) = 1\}$ we get

$$\frac{1}{Q(s)} = \frac{Bs}{2} + \sum_{t=0}^{s-1}\delta(t) = \frac{Bs}{2}\left(1 + \frac{2}{Bs} + \frac{2}{Bs}\sum_{t=0}^{s-1}\delta(t)\right). \tag{27}$$

As we see above $\delta(t) \to 0$ as $t \to \infty$. Hence $(\delta(0) + \cdots + \delta(s-1))/s \to 0$ as $s \to \infty$. Substituting t for s in (27) we have

$$\frac{1}{Q(t)} = \frac{Bt}{2}(1 + o(1)).$$

From this we obtain the assertion of Theorem 3.

The limit behaviour of $Q(t)$ as $t \to \infty$ for critical and subcritical processes is described in Theorems 2 and 3. We now consider the $Q(t)$ in the transitional case $m < 1$, $m \to 1$.

Theorem 4. *Let* $N = 1$, $m < 1$, $m \to 1$, $t \to \infty$, $F'''(1) < \infty$. *Then*

$$Q(t) = m^t \left(1 + \frac{F''(1)}{2} \frac{m^t - 1}{m - 1}\right)^{-1} (1 + o(1)).$$

Proof. Using the Taylor formula for (7) and putting $z = 0$, $Q(t) = 1 - F_t(0)$ we get

$$Q(t+1) = mQ(t) - \frac{F''(1)}{2} Q^2(t) + \frac{C}{6} Q^3(t), \tag{28}$$

where $|C| \leqslant F'''(1)$. Dividing both sides by $mQ(t)Q(t+1)$ we obtain

$$\frac{1}{Q(t+1)} = \frac{1}{mQ(t)} + \frac{F''(1)}{2m} \frac{Q(t)}{Q(t+1)} - \frac{C}{6m} \frac{Q^2(t)}{Q(t+1)}. \tag{29}$$

By Theorem 1 $Q(t) \to 0$ as $t \to \infty$. Therefore from (28) we get

$$\left|\frac{Q(t+1)}{Q(t)} - m\right| \leqslant C_1 Q(t), \quad \left|\frac{Q(t)}{Q(t+1)} - \frac{1}{m}\right| \leqslant C_2 Q(t),$$

where C_1, C_2 denote positive constants. This together with (29) implies that

$$\frac{m^{t+1}}{Q(t+1)} = \frac{m^t}{Q(t)} + \frac{F''(1)}{2m} m^t + h_t, \tag{30}$$

where

$$|h_t| \leqslant C_3 m^t Q(t), \tag{31}$$

C_3 is a positive constant. It is not hard to get from (30) and the equality $Q(0) = 1$ that

$$\frac{m^t}{Q(t)} = 1 + \frac{F''(1)}{2m} \sum_{k=0}^{t-1} m^k + \sum_{k=0}^{t-1} h_k. \tag{32}$$

Since $Q(t) \to 0$ as $t \to \infty$ from (31) we obtain that for any $\varepsilon > 0$ there is such t_0 that

$$\left|\sum_{k=0}^{t-1} h_k\right| \leqslant \sum_{k=0}^{t_0-1} |h_k| + \sum_{k=t_0}^{t-1} C_3 \varepsilon m^k.$$

This implies that

$$\sum_{k=0}^{t-1} h_k = o\left(\sum_{k=0}^{t-1} m^k\right). \tag{33}$$

Since

$$\sum_{k=0}^{t-1} m^k = \frac{m^t - 1}{m - 1}$$

the Theorem 4 follows from (32) and (33).

To progress further we need to consider the asymptotic behaviour of the total progeny of the process G. We denote by ν_N this random variable. We also denote by $f_N(z)$ the generating function of ν_N. Hence

$$f_N(z) = \sum_{k=0}^{\infty} \mathbf{P}\{\nu_N = N + k\} z^{N+k}. \tag{34}$$

Lemma 2. *The equality*

$$f_1(z) = zF(f_1(z)) \tag{35}$$

holds.

Proof. By the total probability formula

$$\mathbf{P}\{\nu_1 = k+1\} = \sum_{i=0}^{\infty} \mathbf{P}\{\mu(1) = i\}\, \mathbf{P}\{\nu_1 = k+1 |\, \mu(1) = i\}. \tag{36}$$

We can consider the particles of the first generation as the initial particles of i branching processes for which the number of direct descendants of a particle has the distribution (1). Therefore

$$\mathbf{P}\{\nu_1 = k+1 |\, \mu(1) = i\} = \sum_{L} \mathbf{P}\{\nu_1 = l_1 + 1\} \ldots \mathbf{P}\{\nu_1 = l_i + 1\},$$

where the summation is over the set

$$L = \{l_1, \ldots, l_i :\ l_1 + \cdots + l_i = k - i\}.$$

Putting this expression into (36) and multiplying both sides by z^{k+1} and summing over k we finally obtain

$$f_1(z) = z \sum_{i=0}^{\infty} \mathbf{P}\{\mu(1)$$

$$= i\} \times \sum_{k=0}^{\infty} \sum_{L} \mathbf{P}\{\nu_1 = l_1 + 1\} z^{l_1+1} \ldots \mathbf{P}\{\nu_1 = l_i + 1\} z^{l_i+1}$$

$$= z \sum_{i=0}^{\infty} p_i \left(\sum_{l=0}^{\infty} \mathbf{P}\{\nu_1 = l+1\} z^{l+1} \right)^i = zF(f_1(z)).$$

Lemma 3. *Let $z \to 1$, $|z| \leqslant 1$, $m = 1$ or $m \to 1$. Then*

$$f_1(z) = 1 - \frac{2\sqrt{1-z}}{\sqrt{2B + w^2} + w}(1 + o(1)),$$

where $w = (1 - zm)/\sqrt{1-z}$, $B = \mathbf{D}\,\mu(1)$.

Proof. From (35) and the equality $F'(1) = m$ we obtain

$$f_1(z) = z(1 + m(f_1(z) - 1) + F''(1)(1 - f_1(z))^2/2), + \varepsilon(z)(1 - f_1(z))^2,$$

where $\varepsilon(z) \to 0$ as $z \to 1$. Solving this equation and choosing a branch of the root such that $0 \leqslant f_1(z) \leqslant 1$ for real z we get the assertion of Lemma 3.

Lemma 4. *Suppose that $N \geqslant 0$, $N + n \geqslant 1$. Then*

$$\mathbf{P}\{\nu_N = N + n\} = \frac{N}{N+n}\mathbf{P}\{\xi_1 + \cdots + \xi_{N+n} = n\},$$

where $\xi_1, \ldots, \xi_{N+n}$ are independent random variables identically distributed by law (1).

Proof. Let $f(N,n) = \mathbf{P}\{\nu_N = N + n\}$ and A_r be the event where the first initial particle has r offspring. By the total probability formula

$$\mathbf{P}\{\nu_N = N + n\} = \sum_{r=0}^{\infty} p_r \, \mathbf{P}\{\nu_N = N + n |\, A_r\}. \tag{37}$$

Since

$$\mathbf{P}\{\nu_N = N + n |\, A_r\} = \mathbf{P}\{\nu_{N+r-1} = N + n - 1\} = f(N + r - 1, n - r),$$

we have

$$f(N,n) = \sum_{r=0}^{\infty} p_r f(N + r - 1, n - r) \tag{38}$$

and as $n \geqslant 1$

$$f(0,n) = \mathbf{P}\{\nu_1 = n + 1 |\, A_0\} = 0, \quad f(1,0) = \mathbf{P}\{\nu_1 = 1\} = p_0. \tag{39}$$

The recurrent relation (38) with the initial conditions (39) uniquely defines $f(N,n)$. To conclude the proof it remains to prove that function

$$f(N,n) = \frac{N}{N+n}\mathbf{P}\{\xi_1 + \cdots + \xi_{N+n} = n\} \tag{40}$$

satisfies the conditions (38), (39). Putting the function (40) on the right side of the equation (38) we obtain

$$\sum_{r=0}^{n} \mathbf{P}\{\xi_{N+n} = r\} \frac{N + r - 1}{N + n - 1} \mathbf{P}\{\xi_1 + \cdots + \xi_{N+n-1} = n - r\}$$

$$= \frac{N-1}{N+n-1}\mathbf{P}\{\xi_1 + \cdots + \xi_{N+n} = n\} + \frac{1}{N+n-1}\sum_{r=0}^{n} r \, \mathbf{P}\{\xi_{N+n} = r\}$$

$$\times \mathbf{P}\{\xi_1 + \cdots + \xi_{N+n-1} = n - r\} = \frac{\mathbf{P}\{\xi_1 + \cdots + \xi_{N+n} = n\}}{N+n-1}$$

$$\times \left[N - 1 + \sum_{r=0}^{n} r \, \mathbf{P}\{\xi_{N+n} = r |\, \xi_1 + \cdots + \xi_{N+n} = n\} \right]$$

$$= \frac{\mathbf{P}\{\xi_1 + \cdots + \xi_{N+n} = n\}}{N+n-1} \left[N - 1 + \mathbf{E}\{\xi_{N+n} |\, \xi_1 + \cdots + \xi_{N+n} = n\} \right].$$

It is easy to see that

$$(N+n)\,\mathbf{E}\{\xi_{N+n}|\ \xi_1+\cdots+\xi_{N+n}=n\}=n,$$

therefore

$$\begin{aligned} f(N,n) &= \sum_{r=0}^{n} p_r \frac{N+r-1}{N+n-1}\,\mathbf{P}\{\xi_1+\cdots+\xi_{N+n-1}=n-r\} \\ &= \frac{N}{N+n}\,\mathbf{P}\{\xi_1+\cdots+\xi_{N+n}=n\}. \end{aligned}$$

Lemma 4 is proved.

We denote by $\mu_r(t)$ the number of particles having exactly r direct descendants in the process G.

Lemma 5. *Let $N \geqslant 0$, $N+n \geqslant 1$. For an integer non-negative $k_0,\ldots,k_n$ such that*

$$\sum_{r=0}^{n} k_r = N+n, \quad \sum_{r=0}^{n} rk_r = n$$

the equalities

$$\mathbf{P}\{\mu_r(t)=k_r,\ r=0,1,\ldots,n;\ \nu_N=N+n\} = \frac{N(N+n-1)!}{k_0!\ldots k_n!}p_0^{k_0}\ldots p_n^{k_n}$$

are true.

Proof. We set

$$f_{k_0,\ldots,k_n}(N,n) = \mathbf{P}\{\mu_r(t)=k_r,\ r=0,1,\ldots,n;\ \nu_N=N+n\}.$$

By the total probability formula

$$f_{k_0,\ldots,k_n}(N,n) = \sum_{r=0}^{n} p_r f_{k_0,\ldots,k_{r-1},k_r-1,k_{r+1},\ldots,k_n}(N+r-1,n-r). \tag{41}$$

It is not hard to see that the recurrent relation (41) with the initial conditions

$$\begin{aligned} &f_{k_0,\ldots,k_n}(0,n)=0 \quad && \text{as } n \geqslant 1, \\ &f_{k_0}(1,0)=p_0 && \text{as } k_0=1 \end{aligned} \tag{42}$$

uniquely defines $f_{k_0,\ldots,k_n}(N,n)$.

It is easy to prove that the function

$$f_{k_0,\ldots,k_n}(N,n) = \frac{N(N+n-1)!}{k_0!\ldots k_n!}p_0^{k_0}\ldots p_n^{k_n}$$

satisfies (41) and (42).

There exists a close relationship between Galton–Watson branching processes and random forests. We consider the set $\mathfrak{F}'_{N,n}$ (see example 1.1) with a uniform probability distribution. Let $\mu_r^{(t)}(f')$ be the number of vertices of height t and outdegree r in a forest from $\mathfrak{F}'_{N,n}$. We denote by G' the Galton–Watson homogeneous branching process with N initial particles and Poisson offspring distribution. Let $\mu_r(t,G')$ be the number of particles at the instant t having exactly r direct descendants, $\nu(G')$ be the total progeny of the process G'. We consider the matrices $||\mu_r^{(t)}(f')||$, $||\mu_r(t,G')||$, $t,r=0,1,\ldots,n$ and the matrix $M=||m_r(t)||$ of the same dimension with entries which are non-negative integers. We will replace f' by $f'(R)$, f'', $f''(R)$ and G' by $G'(R)$, G'', $G''(R)$ when considering respectively $\mathfrak{F}'_{N,n}(R)$, $\mathfrak{F}''_{N,n}$, $\mathfrak{F}''_{N,n}(R)$ (see examples 1.2, 1.3, 1.4) with uniform distributions.

We consider $\mathfrak{F}''_{N,n}$ and the process G'' beginning with N particles such that the number of offspring of any particle ξ'' has a geometric distribution with the parameter α:

$$\mathbf{P}\{\xi''=k\}=\alpha^k(1-\alpha),\quad 0<\alpha<1,\quad k=0,1,\ldots \tag{43}$$

The distributions of the random variables $\mu_r^{(t)}(f'')$ and $\mu_r(t,G'')$ are related in the following way.

Lemma 6. *The equality*

$$\mathbf{P}\{\|\mu_r^{(t)}(f'')\|=M\}=\mathbf{P}\{\|\mu_r(t,G'')\|=M|\,\nu(G'')=N+n\}$$

holds.

Proof. Set $n_t=m_0(t)+\cdots+m_n(t)$. For the equality $\|\mu_r^{(t)}(f'')\|=\|m_r(t)\|$ to have positive probability it is necessary and sufficient that

1) $n_0=N$,
$n_t=m_1(t-1)+2m_2(t-1)+\cdots+nm_n(t-1)$, $t=1,\ldots,n$;
2) if $n_t=0$, then $n_{t+1}=\cdots=n_n=0$;
3) $n_0+n_1+\cdots+n_n=N+n$.

We say that the matrix M is admissible if conditions 1)-3) hold. We consider only the admissible matrix M because, for an inadmissible matrix, both probabilities in the assertion of the Lemma are equal to zero. By Lemma 4 and (43)

$$\begin{aligned}\mathbf{P}\{\nu(G'')=N+n\}&=\frac{N}{N+n}\sum_{k_1+\cdots+k_{N+n}=n}\alpha^{k_1+\cdots+k_{N+n}}(1-\alpha)^{N+n}\\&=\binom{2n+N-1}{n}\frac{N}{N+n}\alpha^n(1-\alpha)^{N+n}.\end{aligned}\tag{44}$$

By definition of conditional probability

$$\begin{aligned}&\mathbf{P}\{\|\mu_r(t,G'')\|=M|\,\nu(G'')=N+n\}\\&=\mathbf{P}\{\|\mu_r(t,G'')\|=M,\ \nu(G'')=N+n\}/\,\mathbf{P}\{\nu(G'')=N+n\}.\end{aligned}\tag{45}$$

It is not hard to see that

$$\begin{aligned}&\mathbf{P}\{\|\mu_r(t,G'')\|=M,\ \nu(G'')=N+n\}\\&=N!n_1!\ldots n_n!\prod_{r,t=0}^{n}\frac{1}{m_r(t)!}\alpha^{rm_r(t)}(1-\alpha)^{m_r(t)}\\&=N!n_1!\ldots n_n!\alpha^n(1-\alpha)^{N+n}\prod_{r,t=0}^{n}\frac{1}{m_r(t)!}.\end{aligned}$$

From this relation and (44), (45) we obtain

$$\mathbf{P}\{\|\mu_r(t,G'')\|=M|\,\nu(G'')=N+n\}=\frac{N!n_1!\ldots n_n!(N+n)}{N\binom{2n+N-1}{n}\prod_{r,t=0}^{n}m_r(t)!}\tag{46}$$

Now let us find the probability $\mathsf{P}\{||\mu_r^{(t)}(f'')|| = M\}$. For the admissible matrix M there exists a forest such that $||\mu_r^{(t)}(f'')|| = M$. By means of this forest we constract all forests for which $||\mu_r^{(t)}(f'')|| = M$. In the stratum of height t there are n_t vertices that are roots of n_t trees. We permute these n_t trees in all $n_t!$ ways and do this in each stratum. The number of forests obtained is, obviously, $N!n_1!\dots n_n!$ It is easily seen that not all these forests are distinct; each forest repeats several times. Identical forests occur for the following reason: $m_r(t)!$ of the permutations of the trees with roots of identical outdegree r in the stratum of height t lead to $m_r(t)!$ repetitions of the same forest. The number of occurrences of each forest is

$$\prod_{r,t=0}^{n} m_r(t)!;$$

therefore, noting by Theorem 1.2 the number of forests in $\mathfrak{F}''_{N,n}$ is equal to

$$\frac{N}{N+n}\binom{2n+N-1}{n},$$

we get the expression on the right-hand side of (46).

For the set $\mathfrak{F}''_{N,n}(R)$ we consider the branching process $G''(R)$ with the offspring distribution

$$\mathsf{P}\{\xi''(R) = k\} = \alpha^k(1-\alpha)/P_1(R,\alpha), \quad k \in R, \tag{47}$$

where $0 < \alpha < 1$,

$$P_1(R,\alpha) = (1-\alpha)\sum_{k\in R} \alpha^k.$$

Lemma 7. *If* $\mathsf{P}\{\nu(G''(R)) = N+n\} > 0$ *then*

$$\mathsf{P}\{||\mu_r^{(t)}(f''(R))|| = M\} = \mathsf{P}\{||\mu_r(t, G''(R))|| = M|\ \nu(G''(R)) = N+n\}.$$

Proof. We denote by $\mu_r(\mathfrak{F}''_{N,n})$ the number of vertices of outdegree r. Introducing the event

$$A_R = \{\mu_r(\mathfrak{F}''_{N,n}) = 0,\ r \notin R\} \tag{48}$$

we note that the probability distribution on $\mathfrak{F}''_{N,n}$ under the condition A_R is uniform. By Lemma 6 and (48)

$$\begin{aligned}&\mathsf{P}\{||\mu_r^{(t)}(f'')|| = M|\ A_R\}\\ &= \mathsf{P}\{||\mu_r(t, G'')|| = M|\ \nu(G'') = N+n,\ A_R(G'')\},\end{aligned}$$

where $A_R(G'') = \{\mu_r(G'') = 0,\ r \notin R\}$. It is clear that

$$\mathsf{P}\{||\mu_r^{(t)}(f'')|| = M|\ A_R\} = \mathsf{P}\{||\mu_r^{(t)}(f''(R))|| = M\}.$$

To conclude the proof, it remains to obtain that

$$\begin{aligned}&\mathsf{P}\{||\mu_r(t, G'')|| = M|\ \nu(G'') = N+n,\ A_R(G'')\}\\ &= \mathsf{P}\{||\mu_r(t, G''(R))|| = M|\ \nu(G''(R)) = N+n\}.\end{aligned} \tag{49}$$

We say that the matrix M is admissible if conditions 1)-3) (see Lemma 6) and condition

4) $m_r(t) = 0$ for $r \notin R$, $t = 0, 1, \ldots, n$,

hold. It is clear that for any inadmissible matrix the equation (49) is valid. Let M be an admissible matrix. By Lemma 5

$$\begin{aligned}&\mathbf{P}\{\mu_r(t, G'') = 0,\ r \in R,\ \nu(G'') = N + n\}\\&= \frac{N(N+n-1)!\alpha^{N+n}(1-\alpha)^n}{P_1^{N+n}(R,\alpha)} \sum_K \frac{1}{k_0! \ldots k_n!},\end{aligned} \tag{50}$$

where the summation is over the set of integers

$$\begin{aligned}K =&\{k_0, k_1, \ldots, k_n :\ k_0 + k_1 + \cdots + k_n\\=&N + n, k_1 + 2k_2 + \cdots + nk_n = n,\ k_r = 0 \text{ for } r \notin R\}.\end{aligned}$$

By Lemma 4

$$\mathbf{P}\{\nu(G''(R)) = N + n\} = \frac{N}{N+n} \mathbf{P}\{\xi_1''(R) + \cdots + \xi_{N+n}''(R) = n\},$$

where $\xi_1''(R), \ldots, \xi_{N+n}''(R)$ are independent random variables with the distribution (47). From (50) we obtain

$$\mathbf{P}\{\nu(G''(R)) = N + n\} = \mathbf{P}\{\nu(G'') = N + n,\ A_R(G'')\}/P_1^{N+n}(R, \alpha). \tag{51}$$

It follows from (47) that

$$\begin{aligned}\mathbf{P}\{\|\mu_r(t, G''(R))\| = M\} &= \frac{\alpha^n(1-\alpha)^{N+n}}{P_1^{N+n}(R,\alpha)} \prod_{t=0}^{n} \frac{n_t!}{m_0(t)! \ldots m_n(t)!}\\&= \mathbf{P}\{\|\mu_r(t, G''(R))\| = M\}/P_1^{N+n}(R, \alpha).\end{aligned}$$

This together with (51) implies (49). The Lemma is proved.

Now we consider the set $\mathfrak{F}'_{N,n}$ (see example 1.1) and the branching process G' such that the number of offspring of one particle ξ' has the Poisson distribution

$$\mathbf{P}\{\xi' = k\} = \frac{\alpha^k e^{-\alpha}}{k!}, \quad \alpha > 0, \quad k = 0, 1, \ldots \tag{52}$$

Lemma 8. *The equality*

$$\mathbf{P}\{\|\mu_r^{(t)}(f')\| = M\} = \mathbf{P}\{\|\mu_r(t, G')\| = M \mid \nu(G') = N + n\}$$

holds.

Proof. We say that the matrix M is admissible if conditions 1)-3) hold (see proof of Lemma 6). For an inadmissible matrix both probabilities are equal to zero. Let M be an admissible matrix. Since the sum of Poisson distributed random variables has Poisson distribution we can obtain from (52) and Lemma 4 by analogy with (44)

$$\mathbf{P}\{\nu(G') = N + n\} = \frac{N(N+n)^{n-1}}{n!} \alpha^n e^{-\alpha(N+n)}. \tag{53}$$

It is easy to see that

$$\begin{aligned}&\mathbf{P}\{\|\mu_r(t, G')\| = M,\ \nu(G') = N + n\}\\&= N! n_1! \ldots n_n! \alpha^n e^{-\alpha(N+n)} \prod_{r,t=0}^{n} (m_r(t)!(r!)^{m_r(t)})^{-1}.\end{aligned}$$

From this and (53) it follows that

$$\mathbf{P}\{||\mu_r(t,G')|| = M|\ \nu(G') = N+n\}$$
$$= \frac{N!n_1!\dots n_n!}{N(N+n)^{n-1}} \prod_{r,t=0}^{n} \frac{1}{m_r(t)!(r!)^{m_r(t)}}. \qquad (54)$$

By analogy with Lemma 6, to find the probability $\mathbf{P}\{||\mu_r^{(t)}(f')|| = M\}$, we permute all subtrees of every stratum of a forest such that $||\mu_r^{(t)}(f')|| = M$. Thus we have $N!n_1!\dots n_n!$ forests but the number of occurrences of each forest is

$$\prod_{r,t=0}^{n} m_r(t)!$$

Also, every vertex with outdegree r leads to $r!$ identical forests. Therefore the number of occurrences of each forest is

$$\prod_{r,t=0}^{n} m_r(t)!(r!)^{m_r(t)}.$$

From this and Theorem 1.1 we get that the probability $\mathbf{P}\{||\mu_r^{(t)}(f')|| = M\}$ is equal to the right-hand side of (54).

It is clear that there exists correspondence between the set $\mathfrak{F}'_{N,n}(R)$ and the branching process $G'(R)$ such that the number of offspring of one particle $\xi'(R)$ has the distribution

$$\mathbf{P}\{\xi'(R) = k\} = \frac{\alpha^k e^{-\alpha}}{k!P_2(R,\alpha)}, \quad k \in R,$$

where

$$P_2(R,\alpha) = e^{-\alpha} \sum_{k\in R} \alpha^k/k!$$

By analogy with Lemma 7 we can prove the following lemma.

Lemma 9. *If* $\mathbf{P}\{\nu(G'(R)) = N+n\} > 0$ *then*

$$\mathbf{P}\{||\mu_r^{(t)}(f'(R))|| = M\} = \mathbf{P}\{||\mu_r(t,G'(R))|| = M|\ \nu(G'(R)) = N+n\}.$$

It is easy to see the similarity of the assertions of Lemmas 6–9. To demonstrate the usefulness of these lemmas we introduce an obvious generalization. We consider a class of forests $\mathfrak{F}_{N,n}$ with N rooted trees and n non-root vertices and the Galton–Watson branching process G beginning with N particles and the offspring distribution (1) such that

$$\mathbf{P}\{||\mu_r^{(t)}(f)|| = M\} = \mathbf{P}\{||\mu_r(t)|| = M|\ \nu_N = N+n\}, \qquad (55)$$

where $\mu_r^{(t)}(f)$ is the number of vertices of height t and of outdegree r in the forest f from $\mathfrak{F}_{N,n}$, $\mu_r(t)$ is the number of particles at instant t having exactly r direct descendants, ν_N is the total progeny of the process G. From Lemmas 6–9 we get that for forests $\mathfrak{F}'_{N,n}$, $\mathfrak{F}'_{N,n}(R)$, $\mathfrak{F}''_{N,n}$, $\mathfrak{F}''_{N,n}(R)$ the relation (55) is true.

By (55), the distribution of the number characteristics of $\mathfrak{F}_{N,n}$, which are functions of $\mu_r^{(t)}(f)$ coincide with those of the corresponding characteristics of the branching process G under the condition $\nu_N = N+n$. Well-known examples of the random variables determined in such a manner are the maximum size of a tree in the forest, the number of trees of a given size, the height of a forest, the number of vertices of a given degree, and the number of vertices of a given height.

For example, let us consider the height of a forest τ. As above, we denote by $\mu(t)$ the number of particles in the t-th generation. From (55) we obtain

$$\mathsf{P}\{\tau < t\} = \mathsf{P}\{\mu(t) = 0|\ \nu_N = N+n\}. \tag{56}$$

Then

$$\mathsf{P}\{\tau < t\} = \frac{\mathsf{P}\{\nu_N = N+n\} - \mathsf{P}\{\mu(t) > 0,\ \nu_N = N+n\}}{\mathsf{P}\{\nu_N = N+n\}}.$$

This relation implies the following assertion

Lemma 10. *The equality*

$$\mathsf{P}\{\tau < t\} = 1 - \mathsf{P}\{\mu(t) > 0\}\,\mathsf{P}\{\nu_N = N+n|\ \mu(t) > 0\}/\,\mathsf{P}\{\nu_N = N+n\}$$

holds.

To prove some results of Chapter 4 we present the probability $\mathsf{P}\{\tau < t\}$ otherwise. The branching process G can be considered as the union of N processes $G^{(1)}, \dots, G^{(N)}$ each of which begins with one particle. Let $\nu^{(i)}$, $i = 1, \dots, N$, stand for the total progenies of the processes $G^{(i)}$. We denote also by $\mu^{(i)}(t)$, $i = 1, \dots, N$, the number of particles in the processes $G^{(i)}$ at the time t. It is clear that

$$\mu(t) = \mu^{(1)}(t) + \cdots + \mu^{(N)}(t), \quad \nu_N = \nu^{(1)} + \cdots + \nu^{(N)}. \tag{57}$$

Let $\nu^{(1)}(t), \dots, \nu^{(N)}(t)$ be independent identically distributed random variables such that

$$\mathsf{P}\{\nu^{(i)}(t) = k\} = \mathsf{P}\{\nu^{(i)} = k|\ \mu^{(i)}(t) = 0\},\ k = 1, 2, \dots,\ i = 1, \dots, N. \tag{58}$$

We also set $\zeta_N^{(t)} = \nu^{(1)}(t) + \cdots + \nu^{(N)}(t)$. From (56) and (58) we obtain the following lemma.

Lemma 11. *The equality*

$$\mathsf{P}\{\tau < t\} = \mathsf{P}\{\mu(t) = 0\}\,\mathsf{P}\{\zeta_N^{(t)} = N+n\}/\,\mathsf{P}\{\nu_N = N+n\}$$

holds.

From Lemmas 2.1, 2.2, 10, 11, examples 2.1, 2.2 we see that the investigation of the asymptotic behaviour of forest characteristics can be reduced to obtaining the limit distributions of the sums of independent random variables. For this we need to use local limit theorems. The next assertion is the classical local limit theorem for sums of independent lattice random variables. A random variable is called *latticed* if all its possible values are presented as $b+kd$, where b is any number, $k = 0, \pm 1, \pm 2, \dots$ and d is a positive number. The number d is called *span* of the distribution of the lattice random variable.

Theorem 5. *Let $N \to \infty$ and independent identically distributed lattice random variables $\xi_1, \xi_2, \dots$ have the mathematical expection m and variance $\sigma^2 > 0$. It is the necessary and sufficient condition for the validity of the relation*

$$\sigma\sqrt{n}\,\mathsf{P}\{\xi_1 + \cdots + \xi_N = Nb + kd\} - \frac{d}{\sqrt{2\pi}} \exp\left\{-\frac{(Nb + kd - Nm)^2}{2N\sigma^2}\right\} \to 0$$

uniformly with respect to k is that the span d is maximal.

We do not give the proof of Theorem 5 for the following reasons. The reader will easily find this proof in many monographs and text-books on probability theory. Furthermore, Theorem 5 very rarely finds applications in the investigations of random forests. In usual situations the sums of independent random variables form array schemes. The known sufficient conditions of the local convergence of such sums do not cover all domains of the variation of parameters. That is why in Chapters 2–4 we derive the proofs of the necessary theorems. The scheme of the proofs of these theorems is similar to the proof of Theorem 5, although the array schemes lead to considerable technical problems. These problems are the main difficulty in obtaining the results of this book.

Theorem 5 sets the conditions of local convergence to normal distribution. However, the need for conditions of convergence for other distributions often arises in study of combinatorial objects by probabilistic methods.

Let $\xi_1, \xi_2, \ldots$ be independent random variables with common distribution function $F_\xi(x)$. If there exist normalizing constants A_N and B_N such that distribution functions of sums $(\xi_1 + \cdots + \xi_N - A_N)/B_N$ weakly converge to some distribution function $G(x)$, then we say that $F_\xi(x)$ is *attracted to* $G(x)$. The set of all such functions is called the *attraction domain* of $G(x)$.

The *stable distribution* with the exponent α is a distribution with the characteristic function

$$\varphi(t) = \exp\{i\gamma t - C|t|^\alpha(1 + i\beta\omega(t,\alpha))\},$$

where $-\infty < \gamma < \infty$, $C > 0$, $|\beta| \leqslant 1$, $0 < \alpha \leqslant 2$, $\omega(t,\alpha) = \operatorname{sign}(\tan(\pi\alpha/2))$ for $\alpha \neq 1$ and $\omega(t,1) = 2\ln t/\pi$.

Theorem 6. *Let $\xi_1, \xi_2, \ldots$ be independent identically distributed lattice random variables, $N \to \infty$, let $g(x)$ be the density of some stable distribution $G(x)$ and let A_N, B_N be some constants. The relation*

$$\sup_k \left| \frac{B_N}{d} \mathbf{P}\{\xi_1 + \cdots + \xi_N = bN + kd\} - g\left(\frac{bN + kd - A_N}{B_N}\right)\right| \to 0$$

is valid if and only if the following conditions hold:

1) *the distribution function of $\xi_1, \xi_2, \ldots$ belongs to the attraction domain of $G(x)$;*
2) *the span d is maximal.*

We do not give the proof of Theorem 6 for the reasons given after Theorem 5. We note that sometimes local limit theorems on large deviations are very useful. For example the next assertion holds.

Theorem 7. *Let $\xi_1, \xi_2, \ldots$ be independent identically distributed lattice random variables with integer values and $d = 1$ and suppose their distribution function belongs to the attraction domain of the stable distribution $G(x)$ with the density $g(x)$ and exponent α, $\alpha \neq 1, 2$. Let $N, k \to \infty$, $kN^{-1/\alpha} \to \infty$,*

$$\mathbf{P}\{\xi_1 = k\} = Ck^{-(1+\alpha)}(1 + o(1)),$$

where C is some positive constant. Then

$$\mathbf{P}\{\xi_1 + \cdots + \xi_N = k\} = N^{-1/\alpha} g(kN^{-1/\alpha})(1 + o(1)).$$

As above, we do not give the proof of Theorem 7 because in the next Chapters we will prove similar theorems for array schemes.

Now we will consider the limit distribution of ν_N when the process G is critical.

Lemma 12. *Let the critical branching process G have the offspring distribution with variance B and maximal span d. Let $r \to \infty$ so that r runs through the natural numbers which are divided by d. Then for any fixed N, k*

$$\mathbf{P}\{\nu_N = N + r + kd\} = \frac{dN(1+\varepsilon(r))}{(N+r+kd)^{3/2}\sqrt{2\pi B}} \exp\left\{-\frac{N^2}{2B(N+r+kd)}\right\},$$

where $\varepsilon(r) \to 0$ and $\varepsilon(r)$ does not depend on N, k.

Proof. By Lemma 4

$$\mathbf{P}\{\nu_N = N + r + kd\} = \frac{N}{N+r+kd}\mathbf{P}\{\xi_1 + \cdots + \xi_{N+r+kd} = r + kd\},$$

where $\xi_1, \ldots, \xi_{N+r+kd}$ are independent random variables with the distribution (1). Using Theorem 5 we get the assertion of Lemma 12.

4. Simply generated forests

Let $\mathfrak{T}$ be a class of plane planted trees. For each $tr \in \mathfrak{T}$ we denote by $\nu(tr)$ the volume of the tree tr. Denote also by $\omega(tr)$ the weight

$$\omega(tr) = \prod_{k=0}^{\infty} \varphi_k^{m_k(tr)}, \tag{1}$$

where φ_k, $k = 0, 1, \ldots$ are non-negative numbers and $m_k(tr)$ is the number of vertices from tr with the outdegree k. Let $\varphi_k^{m_k(tr)} = 1$ if $\varphi_k = m_k(tr) = 0$, $tr \in \mathfrak{F}$ if $\omega(tr) > 0$ and

$$a_n = \sum_{\nu(tr)=n} \omega(tr). \tag{2}$$

We call a family of trees $\mathfrak{T}$ *simply generated* if its generating function

$$a(z) = \sum_{n=0}^{\infty} a_n z^n \tag{3}$$

satisfies the equation

$$a(z) = z\varphi(a(z)), \tag{4}$$

where

$$\varphi(z) = \sum_{k=0}^{\infty} \varphi_k z^k. \tag{5}$$

In Section 1 we considered four examples of forests. It is easily shown that these forests consist of trees from simply generated families of trees. Below in Lemmas 1 and 2 we consider plane planted trees and labelled trees.

Lemma 1. *The set of plane planted trees is a simply generated family of trees.*

Proof. Let $\varphi_k = 1$, $k = 0, 1, \ldots$, then by (5)

$$\varphi(x) = \sum_{k=0}^{\infty} x^k = \frac{1}{1-x}. \tag{6}$$

Using (2) we find that a_n is the number of plane planted trees with n vertices. From this and Lemma 1.1 we obtain that

$$a_n = \binom{2n-2}{n-1} / n, \quad a_0 = 0$$

and the generating function (3) is $zC(z)$ (see (1.4)). Taking into account (6) and the equation $zC^2(z) - C(z) + 1 = 0$ (see proof of Lemma 1.1) we find that (4) is valid. Lemma 1 is proved.

Lemma 2. *The set of rooted labelled trees is a simply generated family of trees.*

Proof. Let $\varphi_k = 1/k!$, $k = 0, 1, \ldots$, then by (5)

$$\varphi(x) = \sum_{k=0}^{\infty} x^k / k! = e^x. \tag{7}$$

From (1) and (2) we obtain

$$a_n (n-1)! = \sum_{\nu(tr)=n} \prod_{k=0}^{n} (n-1)!/(k!)^{m_k(tr)}. \tag{8}$$

It is clear that there exist $(n-1)!$ permutations of $n-1$ vertex numbers in each tree such that $\nu(tr) = n$. But we see that not all of these trees are distinct. Every vertex with the outdegree k gives $k!$ identical trees. Therefore the number of rooted trees with $n-1$ non-root vertices is equal to the right-hand side of the equation (8). From this and Theorem 1.1 we obtain

$$a_n = n^{n-1}/n!, \quad n = 1, 2, \ldots, \quad a_0 = 0.$$

From this we get that the generating function $a(z)$ is the well-known function $\theta(z)$ (see also example 1.1):

$$\theta(z) = \sum_{n=1}^{\infty} \frac{n^{n-1}}{n!} z^n.$$

Since $\theta(z)$ satisfies the equality $\theta(z) = ze^{\theta(z)}$, from (7) we obtain (4). This completes the proof of Lemma 2.

In Section 3 we saw that there exists a correspondence between the considered random trees and Galton–Watson branching processes beginning with one particle. Offspring distributions are geometric for plane planted trees and Poisson for rooted labelled trees. Denote by ξ the number of direct descendants of one particle. Let

$$\mathbf{P}\{\xi = k\} = \varphi_k = \alpha^k (1-\alpha), \quad 0 < \alpha < 1, \quad k = 0, 1, \ldots$$

Using (1) and (5) we get

$$\varphi(z) = (1-\alpha)/(1-\alpha z)$$

and the weight of a tree is the probability of the correspondence realization of the branching process G'' with $N = 1$. It is not hard to see that in this case $a_n = \mathbf{P}\{\nu(G'') = n\}$ and

$$a(z) = \sum_{n=0}^{\infty} \mathbf{P}\{\nu(G'') = n\} z^n.$$

Using this relation and (3.44) we can prove that (4) is true. Furthermore, the equality (4) follows from Lemma 3.2. By analogy, the reader will have no difficulty in proving that (4) is valid for rooted labelled trees and

$$\varphi_k = \mathbf{P}\{\xi = k\} = \frac{\alpha^k}{k!} e^{-\alpha}, \quad \alpha > 0, \quad k = 0, 1, \ldots$$

Now let us check that there is a correspondence between an arbitrary simply generated family of trees $\mathfrak{T}$ and some Galton–Watson branching process with one initial particle. Let

$$p_k = \frac{c^k \varphi_k}{\varphi(c)}, \quad k = 0, 1, \ldots, \tag{9}$$

where c is a non-negative number within the circle of convergence of $\varphi(z)$. We denote by $P(z)$ the generating function

$$P(z) = \sum_{k=0}^{\infty} p_k z^k = \varphi(cz)/\varphi(c). \tag{10}$$

Futher, we can consider the expression

$$p_{tr} = \prod_{k=0}^{\infty} \left(\frac{c^k \varphi_k}{\varphi(c)} \right)^{m_k(tr)} \tag{11}$$

as the weight of the tree $tr \in \mathfrak{T}$. Using this relation and (2) we obtain the sum

$$p_{(n)} = \sum_{\nu(tr)=n} p_{tr} = \frac{c^{n-1}}{\varphi^n(c)} a_n, \quad n = 0, 1, \ldots, \tag{12}$$

and the generating function

$$p(z) = \sum_{n=0}^{\infty} p_{(n)} z^n = \frac{1}{c} a \left(\frac{cz}{\varphi(c)} \right). \tag{13}$$

It is not hard to get from (4) that

$$p(z) = zP(p(z)). \tag{14}$$

Therefore the equations (1)–(5) and equations (9)–(14) give the equivalent ways of characterizing the simply generated family $\mathfrak{T}$. Evidently, we can interpret a tree $t \in \mathfrak{T}$ as the realization of some branching process. Really, if the outdegree of the vertex is k then we say that the correspondence particle has k direct descendants. Also, the linear ordering of the arcs incident to any vertex of a plane planted tree (see Section 1) determines the ordering of items in the sum $\mu(t+1) = \xi_1 + \cdots + \xi_{\mu(t)}$ (see definition of Galton–Watson process in Section 3). It is easy to see that the necessary branching process has the offspring distribution (9) with the generating function (10). If the realization g corresponds to the tree tr then the probability of g is (11) and the distribution of the total progeny is (12).

On the other hand, let $\mathfrak{T}'$ be the set of realizations of a Galton–Watson branching process G with the offspring distribution ξ:

$$p_k(G) = \mathbf{P}\{\xi = k\}, \quad k = 0, 1, \dots$$

and the generating function

$$F_G(z) = \sum_{k=0}^{\infty} p_k(G) z^k.$$

We denote by $\nu(G)$ the total progeny of G and let $f_G(z)$ be the generating function of $\nu(G)$:

$$f_G(z) = \sum_{n=1}^{\infty} \mathbf{P}\{\nu(G) = n\} z^n.$$

Obviously, if $g \in \mathfrak{T}'$, then

$$\mathbf{P}\{G = g\} = \prod_{k=0}^{\infty} (p_k(G))^{m_k(G)},$$

where $m_k(G)$ is the number of particles with k direct descendants. As above, we can see that for any realization $g \in \mathfrak{T}'$ there exists a corresponding plane planted tree tr. By Lemma 3.2 $f_G(z) = zF_G(f_G(z))$ therefore, putting $\varphi_k = p_k(G)$, $k = 0, 1, \dots$, $\varphi(z) = F_G(z)$, $\omega(tr) = \mathbf{P}\{G = g\}$, $a_n = \mathbf{P}\{\nu(G) = n\}$, $a(z) = f_G(z)$ we find that the set $\mathfrak{T}'$ satisfies the definition of a simply generated family of trees.

This reasoning shows that the simply generated family of trees $\mathfrak{T}$ can be regarded as the set of realizations of the Galton–Watson branching process with one initial particle. It is easy to see that the process G induces the probability distribution on $\mathfrak{T}$.

We will now define the class of forests that will be considered in this book. Let $\mathfrak{T}$ be a simply generated family of trees. In the book, we shall discuss the set of forests $\mathfrak{F}_{N,n}$ consisting of N trees from $\mathfrak{T}$ and n non-root vertices.

It is clear that there exists the Galton–Watson branching process G beginning with N particles such that G generates forests with N trees and induces the probability distribution on $\mathfrak{F}_{N,n}$ under the condition $\nu_N = N + n$, where ν_N is the total progeny of G.

Below we will prove the Theorem 1 about the connection of $\mathfrak{F}_{N,n}$ and G. Theorem 1 is the main result of Chapter 1 because in the next Chapters we will use it to prove all results concerning $\mathfrak{F}_{N,n}$.

We denote by $g_{N,n}$ a realization of G with $N + n$ particles. Let $\xi(g_{N,n})$ be some number characteristic of $g_{N,n}$ and let $\xi(G)$ be the corresponding characteristic of G. We denote by $\Delta(g_{N,n})$ the set of forests from $\mathfrak{F}_{N,n}$ which are isomorphic to $g_{N,n}$. Note that the power of $\Delta(g_{N,n})$ can be more than one if we consider a different numbering of vertices. If $f \in \Delta(g_{N,n})$ then we put

$$\xi = \xi(f) = \xi(g_{N,n}).$$

Theorem 1. *The equality*

$$\mathbf{P}\{\xi = x\} = \mathbf{P}\{\xi(G) = x |\ \nu_N = N + n\}$$

holds.

Proof. Let $\chi(A)$ be the indicator of the event A and let $\mathbf{P}\{f\}$ be the probability of f. Then

$$\mathbf{P}\{\xi = x\} = \sum_{g_{N,n}} \sum_{f\in\Delta(g_{N,n})} \mathbf{P}\{f\}\chi(\xi(f) = x)$$

$$= \sum_{g_{N,n}} \chi(\xi(g_{N,n}) = x) \sum_{f\in\Delta(g_{N,n})} \mathbf{P}\{f\} \tag{15}$$

$$= \sum_{g_{N,n}} \chi(\xi(g_{N,n}) = x)\, \mathbf{P}\{G = g_{N,n} |\, \nu_N = N + n\}.$$

It is easy to see that

$$\mathbf{P}\{G = g_{N,n} | \nu_N = N + n\} = \frac{p_0^{m_0} p_1^{m_1} \dots p_n^{m_n}}{\mathbf{P}\{\nu_N = N + n\}}, \tag{16}$$

where p_k is the probability that an arbitrary particle of the process G generates exactly k descendants, $m_0 + m_1 + \dots + m_n = N + n$, $m_1 + 2m_2 + \dots + nm_n = n$ and m_k is the number of particles of $g_{N,n}$ with k direct descendants. It is not hard to prove that (15) and (16) imply the assertion of Theorem 1.

5. Additions and references

Section 1 contains the necessary definitions related to trees and forests. We believe however that the reader would benefit from a more comprehensive knowledge of the basic notions of graph theory and graph enumeration problems. The sources to be recommended in this connection are, for example, the books [26, 27], as well as the classic article [73] by G. Polya. Forests considered in the example 1.1 were studied in [29, 52–54]. Theorem 1.1 was proved in [80]; and Section 1 contains the proof based on Prüfer's method [76]. Forests with constraints on the outdegree of vertices described in the example 1.2 were examined in papers by I. B. Kalugin [30, 32]. Lemma 1.1 was proved in [73], and the proof in Section 1 is based on the derivation of the formula setting Catalan numbers [79]. Forests described in the example 1.3 were investigated in [63, 89]. Theorem 1.2 was proved in co-authorship with V. N. Zemlyachenko in [89]. Forests introduced in the example 1.4 consist of plane planted trees with constraints on the outdegree of vertices; such trees were considered in [62]. The main probability notions used in the books are given in the beginning of Section 2. The random forest is an example of a random element [38]. The generalized allocation scheme was studied in a number of papers by V. F. Kolchin, and is described in most detail in the monographs [38, 43]. The connection between the generalized allocation scheme and random forests from $\mathfrak{F}'_{N,n}$ considered in the example 2.1 was demonstrated in [52, 53]. Section 3 describes the connection between some classes of forests and Galton–Watson branching processes beginning with N particles. These results are a naturally-following generalization of the theorems on random trees obtained in [38] using branching processes beginning with one particle. Descriptions of the main notions of the branching process theory used in the book can be found in [38, 81]. Theorems 3.1, 3.2 and Lemma 3.1 were proved in [81]. Corollary 3.1 was mentioned in [57], and presented in more detail in [63]. Theorems 3.3 and 3.4 were proved in [81], Lemma 3.2 and Lemma 3.3 for the case $m = 1$ in [38]. Lemma 3.4 was obtained in [88], and the proofs of Lemmas 3.4 and 3.5 in our book follow [38]. The article [33] deals with the branching process distributions under the condition that the total number of particles in the process is known. The connection between plane planted trees and Galton–Watson branching processes with the geometric distribution of the number of offspring of one particle was mentioned in [61], and Lemma 3.6 was proved in [89]. Lemma 3.8 is given in [57], Lemma 3.9 in [30]. Lemmas 3.10 and 3.11 relying on the relation (3.55) were proved in [64]. Theorem 3.5 belongs to B. V. Gnedenko [22]. The application of local limit theorems to sums of independent random variables in the study of combinatorial objects is discussed in detail in Section 1.4 of the book [38] by V. F. Kolchin. In this text we will note only that the known sufficient conditions for convergence in an array scheme (e. g. see [50]) do not cover all domains of variation of the parameters needed to obtain results for random forests. Theorem 3.6 was proved in [28], formulation of Theorem 3.7 was given in [86] as well as the idea of the proof of this Theorem. Lemma 3.12 is a revision of Lemma 2.1.4 from [38].

The notion of a simply generated family of trees was introduced in [49]. Any family of this kind can be considered as a set of realizations of some Galton–Watson process — a fact demonstrated in [4]. Thus, using the properties of branching processes, one can study various characteristics of simply generated trees (e. g. see [16–18]).

Theorem 4.1 was proved in [13]. An example of using Theorem 4.1 for the study of random forests is the article [29], which considered distributions of the outdegree of vertices in a forest from the set $\mathfrak{F}'_{N,n}$ (see example 1.1) on which uniform probability distribution had been set. Let μ_r be the number of vertices with the outdegree r in a forest from $\mathfrak{F}'_{N,n}$, $r = 0, 1, \ldots, n$, $\mu_r(G_r)$ be the number of particles of the corresponding branching process G_r with exactly r direct descendants. It follows from Theorem 4.1 that

$$\begin{aligned} &\mathbf{P}\{\mu_r = k_r,\ r = 0, 1, \ldots, n\} \\ &\quad = \mathbf{P}\{\mu_r(G) = k_r,\ r = 0, 1, \ldots, n \mid \nu_N = N + n\}. \end{aligned} \tag{1}$$

It was demonstrated in [29] that

$$\begin{aligned} &\mathbf{P}\{\mu_r(G) = k_r,\ r = 0, 1, \ldots, n \mid \nu_N = N + n\} \\ &\mathbf{P}\{\mu_r(n, N + n) = k_r,\ r = 0, 1, \ldots, n\}, \end{aligned}$$

where $\mu_r(n, N{+}n)$ is the number of cells with r particles under equiprobable allocation of the particles to cells. From this and (1) it follows that the limit distributions of μ_r coincide with the limit distributions of $\mu_r(n, N + n)$, the full description of which is given in the book by V. F. Kolchin, B. A. Sevastyanov and V. P. Chistyakov [43].

Important results for random forests of labelled non-rooted trees were obtained by V. E. Britikov [7–10].

CHAPTER 2

THE MAXIMUM SIZE OF A TREE IN A RANDOM FOREST

1. Problem statement and summary of results

We shall consider the class of forests $\mathfrak{F}_{N,n}$ that was introduced in Section 1.4. Let $\nu_1(\mathfrak{F}), \dots, \nu_N(\mathfrak{F})$ be the sizes of trees with roots $1, \dots, N$ in a forest from $\mathfrak{F}_{N,n}$. In Chapter 1 we saw that there exists correspondence between $\mathfrak{F}_{N,n}$ and some Galton–Watson branching process G beginning with N particles. The process G consists of N independent processes $G^{(1)}, \dots, G^{(N)}$ with one initial particle. Let ξ be the auxiliary random variable with the probability distribution

$$\mathbf{P}\{\xi = k\} = p_k, \quad k = 0, 1, 2, \dots, \tag{1}$$

maximum span d and generating function

$$F(z) = \sum_{k=0}^{\infty} p_k z^k. \tag{2}$$

We assume also that the set of values ξ with non-zero probability includes null and differs from $\{0, 1\}$. Let $\mathbf{E}\xi = 1$, $\mathbf{D}\xi = B$ and suppose that the number of offspring of a particle in the process G has the distribution

$$p_k(\lambda) = \lambda^k p_k / F(\lambda), \quad k = 0, 1, 2, \dots, \tag{3}$$

where $0 < \lambda \leqslant 1$. The generating function of the distribution (3) is

$$F_\lambda(z) = \sum_{k=0}^{\infty} p_k(\lambda) z^k = \sum_{k=0}^{\infty} (\lambda z)^k p_k / F(\lambda). \tag{4}$$

Let $\nu^{(i)}$ stand for the total progenies of the processes $G^{(i)}$, $i = 1, \dots, N$, $\nu_N = \nu^{(1)} + \dots + \nu^{(N)}$. By Theorem 1.4.1 if $\mathbf{P}\{\nu_N = N + n\} > 0$ then

$$\begin{aligned} &\mathbf{P}\{\nu_1(\mathfrak{F}) = k_1, \dots, \nu_N(\mathfrak{F}) = k_N\} \\ &= \mathbf{P}\{\nu^{(1)} = k_1, \dots, \nu^{(N)} = k_N \mid \nu_N = N + n\}. \end{aligned} \tag{5}$$

We denote by $\nu_{[r]}^{(1)}, \dots, \nu_{[r]}^{(N)}$ independent identically distributed random variables such that

$$\mathbf{P}\{\nu_{[r]}^{(i)} = k\} = \mathbf{P}\{\nu^{(i)} = k \mid \nu^{(i)} \leqslant r + 1\}, \quad i = 1, \dots, N, \quad k = 1, 2, \dots \tag{6}$$

We set also $\nu_N^{(r)} = \nu_{[r]}^{(1)} + \dots + \nu_{[r]}^{(N)}$, $P_r = \mathbf{P}\{\nu^{(1)} > r + 1\}$; therefore

$$P_r = \sum_{k=1}^{\infty} \mathbf{P}\{\nu^{(1)} = r + kd + 1\}. \tag{7}$$

Let η be the maximum tree size in a forest from $\mathfrak{F}_{N,n}$. Hence

$$\eta = \max_{1 \leqslant i \leqslant N} \nu_i(\mathfrak{F}).$$

It is easy to see that (5) is an example of (1.2.1) if $\xi_i = \nu^{(i)} - 1$, $i = 1, \dots, N$. Therefore the conditions of the generalized allocation scheme are valid (see Section 1.2) and from Lemma 1.2.2 we obtain the next assertion.

Lemma 1. *If* $\mathsf{P}\{\nu_N = N + n\} > 0$ *then*

$$\mathsf{P}\{\eta \leqslant r\} = (1 - P_r)^N \frac{\mathsf{P}\{\nu_N^{(r)} = N + n\}}{\mathsf{P}\{\nu_N = N + n\}}. \tag{8}$$

Lemma 1 shows that in order to obtain the limit distributions of η it suffices to consider the asymptotic behaviour of the binomial $(1 - P_r)^N$ and the probabilities $\mathsf{P}\{\nu_N^{(r)} = N + n\}$, $\mathsf{P}\{\nu_N = N + n\}$. We will get these results in Sections 2–5 and in Section 6 we will prove the following theorems 1–5.

We denote by j the least positive integer such that $p_j > 0$ and by l the least natural number not divided by j for which $p_{j+l} > 0$; if such l does not exist, we put $l = 0$.

Theorem 1. *Let $N, n \to \infty$ in such a way that n takes values which are divided by d, $n/N \to 0$ and let $\lambda = \lambda(N, n)$ be determined by the relation*

$$\frac{\lambda F'(\lambda)}{F(\lambda)} = \frac{n}{N + n}. \tag{9}$$

Let

$$N\,\mathsf{P}\{\nu^{(1)} = r + 1\} \to \infty, \quad N\,\mathsf{P}\{\nu^{(1)} = r + s + 1\} \to \gamma,$$

where γ is a non-negative constant and natural r and s are devided by d and satisfy one of the conditions

$$r \to \infty, \quad s = d,$$

or r is fixed, $r \geqslant j + l$, the maximum span of the distribution of $\nu_{[r]}^{(1)}$ is d, $\mathsf{P}\{\nu^{(1)} = r + s + 1\} > 0$, $\mathsf{P}\{\nu^{(1)} = r + i + 1\} = 0$, $0 < i < s$. Then

$$\mathsf{P}\{\eta = r + 1\} \to e^{-\gamma}, \quad \mathsf{P}\{\eta = r + s + 1\} = 1 - e^{-\gamma}.$$

Remark. Some of the sufficient conditions for $r = r(N, n)$ in Theorem 1 we will obtain in Section 6.

Theorem 2. *Let $N, n \to \infty$ in such a way that n takes values which are divided by d and $n/N \to b$, where b is a positive constatnt. Let $\lambda = \lambda(N, n)$ be given by (9) and*

$$\alpha = (\lambda_b / F(\lambda_b))^d, \tag{10}$$

where λ_b is the solution of the equation

$$\lambda F'(\lambda)/F(\lambda) = b/(b + 1).$$

If $r = r(N, n)$ running throught the values which are divided by d is such that

$$\frac{N}{F(\lambda)} \left(\frac{\lambda}{F(\lambda)}\right)^r \frac{d}{r^{3/2}\sqrt{2\pi B}} \to \gamma, \tag{11}$$

where γ is a positive constant, then for any fixed $k = 0, \pm1, \pm2 \dots$

$$\mathbf{P}\{\eta \leqslant r + kd + 1\} \to \exp\{-\gamma\alpha^{k+1}(1-\alpha)^{-1}\}.$$

Theorem 3. *Let $F'''(1) < \infty$, $N, n \to \infty$ in such a way that n takes values which are divided by d, $n/N \to \infty$, $n/N^2 \to 0$. Let $\lambda = \lambda(N, n)$ be defined by (9). Then*

$$\mathbf{P}\{\beta\eta - u \leqslant z\} \to e^{-e^{-z}},$$

where

$$\beta = \beta(\lambda) = -\ln(\lambda/F(\lambda)), \tag{12}$$

and $u = u(\lambda)$ is chosen so that

$$N\beta^{1/2}u^{-3/2}e^{-u} = \sqrt{2\pi B}. \tag{13}$$

Theorem 4. *Let $N, n \to \infty$ in such a way that n takes values which are divided by d, $Bn/N^2 \to \gamma$, where γ is a positive constant. Then*

$$\mathbf{P}\{\eta/n \leqslant z\} \to \gamma^{3/2}\exp\{1/(2\gamma)\}\sum_{k=0}^{\infty}(-1)^k(k!)^{-1}I_k(\gamma z, \gamma),$$

where

$$I_0(u,v) = (v^3\exp\{1/v\})^{-1/2},$$

$$I_k(u,v) = \int\limits_{X_k(u,v)} \frac{\exp\{-1/(2(v - x_1 - \dots - x_k))\}dx_1 \dots dx_k}{(2\pi)^{k/2}(x_1 \dots x_k(v - x_1 - \dots - x_k))^{3/2}},$$

$$X_k(u,v) = \{x_i \geqslant u,\ i = 1, \dots, k,\ x_1 + \dots + x_k \leqslant v\}, \quad k = 1, 2, \dots$$

Theorem 5. *Let $n \to \infty$ in such a way that n takes values which are divided by d, $n/N^2 \to \infty$. Then for any fixed positive z*

$$\mathbf{P}\left\{\frac{n-\eta}{N^2}B \leqslant z\right\} \to \frac{1}{\sqrt{2\pi}}\int\limits_0^z y^{-3/2}\exp\{-1/(2y)\}dy.$$

2. Asymptotics of NP_r

In this Section we shall consider the limit behaviour of NP_r. The results proved below fully describe the asymptotics of the binomial $(1 - P_r)^N$ which stands on the right-hand side of (1.8). The behaviour of the parameter $\lambda = \lambda(N, n)$ (see (1.9)) is studied in Lemma 1. Let m be the mathematical expectation of the distribution (1.3). It is easy to see that

$$m = m(\lambda) = \lambda F'(\lambda)/F(\lambda). \tag{1}$$

Lemma 1. *There exists a unique solution λ of the equation (1.9) which satisfies the inequalities $0 \leqslant \lambda \leqslant 1$. For this solution the next assertions are valid as $N, n \to \infty$:*

1) *if $n/N \to 0$ and j is the least positive integer such, that $p_j > 0$, then*

$$\lambda = \left(\frac{n}{N}\frac{p_0}{jp_j}\right)^{1/j}(1 + o(1));$$

2) *if* $0 < C_1 \leqslant n/N \leqslant C_2 < \infty$, *then* $0 < C_3 \leqslant \lambda \leqslant C_4 < 1$;
3) *if* $n/N \to \infty$, *then* $\lambda \to 1$.

Proof. Since $p_0 > 0$ it follows that, for any $\lambda \geqslant 0$, the inequalities $F(\lambda) \geqslant p_0 > 0$ are true. If we combine this with (1.1), (1) and the condition $\mathbf{E}\xi = 1$ we get $m(0) = 0$, $m(1) = 1$. Using (1) we obtain

$$m'(\lambda) = F'(\lambda)[F(\lambda) - \lambda F'(\lambda)]F^{-2}(\lambda). \tag{2}$$

By (1.9), $\lambda F'(\lambda) < F(\lambda)$ so that (2) implies $m'(\lambda) > 0$. We see that the function $m(\lambda)$ monotonously increases as $0 \leqslant \lambda \leqslant 1$. Therefore the solution of (1.9) is unique. If $n/N \to 0$ then $m(\lambda) \to 0$ and using (1) we obtain $\lambda \to 0$. Then we get the first assertion from (1), (1.9) and evident expressions of $F(\lambda)$ and $F'(\lambda)$. The second and third assertions we also get from (1) and (1.9).

Lemma 2. *Under the conditions of Theorem* 1.1, $NP_{r-d} \to \infty$, $NP_{r+k} \to \gamma$, $k = 0, d, 2d, \ldots, s-d$.

Proof. It is easy to see that for $k = 0, d, 2d, \ldots, s-d$

$$\begin{aligned} P_{r+k} &= \sum_{i=0}^{\infty} \mathbf{P}\{\nu^{(1)} = r+s+id+1\} \\ &= \mathbf{P}\{\nu^{(1)} = r+s+1\}\left(1 + \sum_{i=1}^{\infty} \frac{\mathbf{P}\{\nu^{(1)} = r+s+id+1\}}{\mathbf{P}\{\nu^{(1)} = r+s+1\}}\right). \end{aligned} \tag{3}$$

By Lemma 1.3.4

$$\mathbf{P}\{\nu^{(1)} = n+1\} = (n+1)^{-1}\,\mathbf{P}\{\xi_1(\lambda) + \cdots + \xi_{n+1}(\lambda) = n\},$$

where $\xi_1(\lambda), \ldots, \xi_{n+1}(\lambda)$ are independent random variables with the distribution (1.3). From this it follows that

$$\begin{aligned} \mathbf{P}\{\nu^{(1)} = n+1\} &= \frac{\lambda^n}{F^{n+1}(\lambda)}\,\mathbf{P}\{\xi_1(1) + \cdots + \xi_{n+1}(1) = n\} \\ &= \frac{\lambda^n}{F^{n+1}(\lambda)}\,\mathbf{P}\{\nu_*^{(1)} = n+1\}, \end{aligned} \tag{4}$$

where $\xi_1(1), \ldots, \xi_{n+1}(1)$ are independent random variables with the distribution (1.1) and $\nu_*^{(1)}$ is the total progeny of the critical Galton–Watson process beginning with one particle for which the number of offspring of a particle has the distribution (1.1). From (3), (4) we obtain

$$\begin{aligned} P_{r+k} &= \mathbf{P}\{\nu^{(1)} = r+s+1\} \\ &\times \left(1 + \sum_{i=1}^{\infty}\left(\frac{\lambda}{F(\lambda)}\right)^{id} \frac{\mathbf{P}\{\nu_*^{(1)} = r+s+id+1\}}{\mathbf{P}\{\nu_*^{(1)} = r+s+1\}}\right). \end{aligned} \tag{5}$$

Let $r \to \infty$. By Lemma 1.3.12 uniformly with respect to natural l

$$\mathbf{P}\{\nu_*^{(1)} = r+ld+1\} = r^{-3/2}(2\pi B)^{-1/2}(1+o(1)). \tag{6}$$

From Lemma 1 it follows that, under the conditions of Lemma 2, $\lambda \to 0$; hence from (5) and (6) we get

$$P_{r+k} = \mathbf{P}\{\nu^{(1)} = r + s + 1\}(1 + o(1)). \tag{7}$$

If r is fixed then (5) implies (7). By analogy we can get that $P_{r-d} = \mathbf{P}\{\nu^{(1)} = r + 1\}(1 + o(1))$. From this and (7) we obtain the assertion of Lemma 2.

Lemma 3. *Under the conditions of Theorem 1.2 for any fixed integer k*

$$NP_{r+kd} \to \gamma\alpha^{k+1}(1-\alpha)^{-1}.$$

Proof. By Lemma 1 there exists a unique solution of the equation (1.9). It is easy to prove that

$$\alpha = (\lambda_b/F(\lambda_b))^d = \lim_{N,n\to\infty} (\lambda/F(\lambda))^d. \tag{8}$$

Note that

$$NP_{r+kd} = N\,\mathbf{P}\{\nu^{(1)} = r+1\} \sum_{i=0}^{\infty} \frac{\mathbf{P}\{\nu^{(1)} = r + (k+i+1)d + 1\}}{\mathbf{P}\{\nu^{(1)} = r+1\}}. \tag{9}$$

From (4) we see that for any natural i

$$\begin{aligned} &\mathbf{P}\{\nu^{(1)} = r + (k+i+1)d + 1\}/\,\mathbf{P}\{\nu^{(1)} = r+1\} \\ &= \left(\frac{\lambda}{F(\lambda)}\right)^{(k+i+1)d} \frac{\mathbf{P}\{\nu_*^{(1)} = r + (k+i+1)d + 1\}}{\mathbf{P}\{\nu_*^{(1)} = r+1\}}. \end{aligned} \tag{10}$$

Using Lemma 1.3.12 to estimate the probabilities standing on the right-hand side of the last equation we obtain

$$\frac{\mathbf{P}\{\nu^{(1)} = r + (k+i+1)d + 1\}}{\mathbf{P}\{\nu^{(1)} = r+1\}} = \left(\frac{\lambda}{F(\lambda)}\right)^{(k+i+1)d} \left(1 + O\left(\frac{i}{r}\right)\right). \tag{11}$$

From Lemma 1.3.12 and (4), (1.11) we find that

$$N\,\mathbf{P}\{\nu^{(1)} = r+1\} \to \gamma. \tag{12}$$

By (1.9), $\lambda F'(\lambda) < F(\lambda)$; therefore the function $\lambda/F(\lambda)$ monotonically increases from 0 to 1 as $0 \leqslant \lambda \leqslant 1$. From this and (8) we obtain $\alpha < 1$. Using this inequality and (8)–(12) we see that

$$NP_{r+kd} \to \gamma \sum_{i=0}^{\infty} (\lambda_b/F(\lambda_b))^{(k+i+1)d} = \gamma\alpha^{k+1}(1-\alpha)^{-1}.$$

Lemma 3 is proved.

Corollary 1. *Let $N, n \to \infty$ such that $n/N \to b$, where b is some positive constant. Let $\lambda = \lambda(N,n)$ be given by (1.9) and α be defined by (1.10). If $r = r(N,n)$ running through the values which are divided by d is such that $NP_r \to \gamma$, where γ is a positive constant, then*

$$N\,\mathbf{P}\{\nu^{(1)} = r+1\} \to \gamma\alpha^{-1}(1-\alpha).$$

Proof. Obviously, the assertion of the Corollary follows from (9), as does the relation

$$\mathbf{P}\{\nu^{(1)} = r + id + 1\} / \mathbf{P}\{\nu^{(1)} = r + 1\} = \alpha^i(1 + O(i/r)), \tag{13}$$

which is true as $r \to \infty$ with fixed natural i.

To prove some results below we will use the next Lemma 4. It is not hard to derive (for example by L'Hospital rule) such assertion.

Lemma 4. *Let $x \to \infty$. Then for any fixed h*

$$\int_x^\infty y^h e^{-y} dy = x^h e^{-x}(1 + o(1)).$$

Lemma 5. *Let $N, n \to \infty$ in such a way that $n/N \to \infty$, $n/N^2 \to 0$, $\lambda = \lambda(N,n)$ be given by* (1.9). *Let r be divided by d and $r = (u+z)/\beta + O(1)$, where z is fixed, u and β are defined by* (1.13) *and* (1.12) *respectively. Then $NP_r \to e^{-z}$.*

Proof. From (1.6) and (4) it follows that

$$P_r = \frac{1}{F(\lambda)} \sum_{k=1}^{\infty} \left(\frac{\lambda}{F(\lambda)}\right)^{r+kd} \mathbf{P}\{\nu_*^{(1)} = r + kd + 1\}. \tag{14}$$

By Lemma 1, $\lambda \to 1$; therefore $F(\lambda) \to 1$. From this relation, (14) and Lemma 1.3.12 we obtain

$$\begin{aligned} P_r &= \frac{d(1+o(1))}{\sqrt{2\pi B}} \sum_{k=1}^{\infty} \left(\frac{\lambda}{F(\lambda)}\right)^{r+kd} \frac{1}{(r+kd)^{3/2}} \\ &= \frac{d(1+o(1))}{\sqrt{2\pi B}} \sum_{k=1}^{\infty} (r+kd)^{-3/2} \exp\{-(r+kd)\beta\}. \end{aligned} \tag{15}$$

It is not hard to see that

$$\begin{aligned} &\int_1^\infty (r+yd)^{-3/2} \exp\{-(r+yd)\beta\} dy \\ &< \sum_{k=1}^{\infty} (r+kd)^{-3/2} \exp\{-(r+kd)\beta\} \\ &< \int_0^\infty (r+yd)^{-3/2} \exp\{-(r+yd)\beta\} dy. \end{aligned} \tag{16}$$

Putting $v = r + yd$ we get

$$\begin{aligned} d^{-1} \int_{r+d}^\infty v^{-3/2} e^{-v\beta} dv &< \sum_{k=1}^{\infty} (r+kd)^{-3/2} \exp\{-(r+kd)\beta\} \\ &< d^{-1} \int_r^\infty v^{-3/2} e^{-v\beta} dv. \end{aligned} \tag{17}$$

From (15) and (17) we find

$$\begin{aligned} P_r &= (2\pi B)^{-1/2} \int_r^\infty v^{-3/2} e^{-v\beta} dv + O\left(\int_r^{r+d} v^{-3/2} e^{-v\beta} dv \right) \\ &= (1+o(1)) \sqrt{\frac{\beta}{2\pi B}} \int_{r\beta}^\infty v^{-3/2} e^{-v} dv + O\left(\int_{r\beta}^{(r+d)\beta} v^{-3/2} e^{-v} dv \right). \end{aligned} \tag{18}$$

For the generating function $F(\lambda)$ there exists the left derivative $F''(1) = B$. Expanding $F(\lambda)$ in series in the neighbourhood of the point $\lambda = 1$ and using (1.12) and equation $F'(1) = 1$ we obtain

$$\beta = (B(1-\lambda)^2/2)(1+o(1)). \tag{19}$$

Using the condition $N^2/n \to \infty$ and expanding $F(\lambda)$ and $F'(\lambda)$ in series by virtue of (1.9) we get $N(1-\lambda) \to \infty$. Substituting (19) for (1.13) we see that

$$u = \ln N(1-\lambda) - (3/2) \ln \ln N(1-\lambda) - \ln(2\sqrt{\pi}) + o(1)$$

and $u \to \infty$. Therefore, $r\beta \to \infty$ and using Lemma 4 and (18) we obtain

$$P_r = \sqrt{\frac{\beta}{2\pi B}} u^{-3/2} e^{-(u+z)} (1+o(1));$$

hence, by virtue of (1.13)

$$NP_r = \frac{N\beta^{1/2}}{u^{3/2}\sqrt{2\pi B}} e^{-(u+z)} = e^{-z}(1+o(1)).$$

Lemma 5 is proved.

Lemma 6. *Let $\lambda = 1$, $N, n \to \infty$ in such a way that $Bn/N^2 \to \gamma$, and let r be divided by d and $r = nz + O(1)$, where γ, z are positive constants. Then*

$$NP_r \to (2\pi)^{-1/2} \int_{\gamma z}^\infty y^{-3/2} dy.$$

Proof. By Lemma 1.3.12

$$P_r = \sum_{k=1}^\infty \mathbf{P}\{\nu^{(1)} = r + kd + 1\} = d(2\pi B)^{-1/2}(1+o(1)) \sum_{k=1}^\infty (r+kd)^{-3/2}.$$

Putting $y = (r+kd)/N^2$ and changing the summation for integration we obtain

$$\begin{aligned} P_r &= (N\sqrt{2\pi B})^{-1} \int_{\gamma z/B}^\infty y^{-3/2} dy + o(N^{-1}) \\ &= (N\sqrt{2\pi})^{-1} \int_{\gamma z}^\infty y^{-3/2} dy + o(N^{-1}); \end{aligned} \tag{20}$$

the Lemma is thus proved.

The relation (20) implies the next assertion.

Lemma 7. *Let $\lambda = 1$, $n \to \infty$ in such a way that $n/N^2 \to \infty$ and let r be divided by d and $r = n(1 + o(1))$. Then $NP_r \to 0$.*

3. The limit behaviour of the total progeny of the branching process

In this Section we shall consider the limit behaviour of the random variable ν_N which is equal to the total progeny over the whole period of evolution of the process G. We will obtain the limit expressions of the probability $\mathbf{P}\{\nu_N = N + n\}$ for various domains of variation of N and n. In Lemmas 1 and 2 below we consider the asymptotics of ν_N as $N, n \to \infty$ such that $n/N^2 \to 0$, $\lambda < 1$ and Lemmas 3 and 4 are intended to clarify the limit behaviour of ν_N as $n/N^2 \geqslant C > 0$, $\lambda = 1$.

We denote by $\varphi(u)$ the characteristic function of $\nu^{(1)}$. From Lemma 1.3.2 we obtain

$$\varphi(u) = e^{iu} F_\lambda(\varphi(u)). \tag{1}$$

Let $\lambda < 1$, $a = \mathbf{E}\,\nu^{(1)}$, $\sigma^2 = \mathbf{D}\,\nu^{(1)}$. Using the equation (1) it is not hard to find $\varphi'(u)$, $\varphi''(u)$ and get the equalities

$$a = \mathbf{E}\,\nu^{(1)} = 1/(1-m), \quad \sigma^2 = \mathbf{D}\,\nu^{(1)} = B_\lambda/(1-m)^3, \tag{2}$$

where m and B_λ denote the mathematical expectation and the variance of the distribution (1.3) respectively.

Let $\varphi_N(u)$ be the characteristic function of the random variable $(\nu_N - N - n)/(\sigma\sqrt{N})$. The next assertion is valid.

Lemma 1. *Let j be the least positive integer such that $p_j > 0$. Let $N, n \to \infty$ such that $n/N^2 \to 0$, $N\lambda^j \to \infty$, where $\lambda = \lambda(N, n)$ is given by (1.9) and $F'''(1) < \infty$ if $n/N \to \infty$. Then*

$$\varphi_N(u) \to e^{-u^2/2}$$

uniformly in u in any finite interval.

Proof. To prove Lemma 1 it is important to know the evident expression of the third derivative of $\ln\varphi(u)$. Using (1) we can prove that

$$\begin{aligned}(\ln\varphi(u))'''_u &= -i(1 - e^{iu}(F_\lambda(\varphi(u)))'_\varphi)^{-5} e^{iu} \\ &\times\{(F_\lambda(\varphi(u)))'''_\varphi \varphi^2(u) - e^{iu}(F_\lambda(\varphi(u)))''_\varphi (F_\lambda(\varphi(u)))'_\varphi \varphi^2(u) \\ &-e^{iu}((F_\lambda(\varphi(u)))'_\varphi)^2 + 3(F_\lambda(\varphi(u)))''_\varphi \varphi(u) \\ &+e^{3iu}((F_\lambda(\varphi(u)))'_\varphi)^4 - 3e^{2iu}(F_\lambda(\varphi(u)))''_\varphi((F_\lambda(\varphi(u)))'_\varphi)^2\varphi(u) \\ &+3e^{iu}(F_\lambda(\varphi(u)))''_\varphi \varphi^2(u) + (F_\lambda(\varphi(u)))'_\varphi - e^{2iu}((F_\lambda(\varphi(u)))'_\varphi)^3\}.\end{aligned} \tag{3}$$

If $n/N \geqslant C_1 > 0$, then from (2) and (1.9) it follows that $\sigma^2 \geqslant C_2 > 0$. If $n/N \to 0$, then by Lemma 2.1 $\lambda \to 0$ and from (1.3), (1.9), (2.1) we obtain

$$m \to 0, \quad B_\lambda = O(\lambda^j). \tag{4}$$

Therefore, from (2) we get

$$\sigma\sqrt{N} \to \infty. \tag{5}$$

Since $\nu_N = \nu^{(1)} + \cdots + \nu^{(N)}$ it follows that

$$\varphi_N(u) = \exp\left\{-\frac{i(N+n)u}{\sigma\sqrt{N}}\right\}\varphi^N\left(\frac{u}{\sigma\sqrt{N}}\right).$$

Hence

$$\ln\varphi_N(u) = -\frac{i(N+n)u}{\sigma\sqrt{N}} + N\ln\varphi\left(\frac{u}{\sigma\sqrt{N}}\right). \tag{6}$$

The next expansion for sufficiently small u is valid:

$$\begin{aligned}\ln\varphi(u) =& u(\ln\varphi(u))'_{u=0} + \frac{u^2}{2}(\ln\varphi(u))''_{u=0} + \frac{u^3}{3!}Q(u)\\ =& iua - \frac{u^2\sigma^2}{2} + \frac{u^3}{6}Q(u),\end{aligned} \tag{7}$$

$$|Q(u)| \leqslant 2\max_{|\tau|\leqslant|u|}|(\ln\varphi(\tau))'''_\tau|. \tag{8}$$

From (5)–(8) it follows that for sufficiently large N, n and any fixed u

$$\ln\varphi_N(u) = -\frac{u^2}{2} + \frac{u^3}{6\sigma^3\sqrt{N}}Q\left(\frac{u}{\sigma\sqrt{N}}\right). \tag{9}$$

Let us prove that the second term on the right-hand side of the equation (9) tends to zero. From (8) it follows that

$$\left|\frac{u^3}{6\sigma\sqrt{N}}Q\left(\frac{u}{\sigma\sqrt{N}}\right)\right| \leqslant \frac{|u|^3}{3}Q_1(u), \tag{10}$$

where

$$Q_1(u) = \max_{|\tau|\leqslant|u/(\sigma\sqrt{N})|}|(\ln\varphi(\tau))'''_\tau|/(\sigma^3\sqrt{N}). \tag{11}$$

Since $|\varphi(u)| \leqslant 1$ we see that $|e^{iu}(F_\lambda(\varphi(u)))'_\varphi| \leqslant m$; therefore

$$|1 - e^{iu}(F_\lambda(\varphi(u)))'_\varphi| \geqslant 1 - m = N/(N+n). \tag{12}$$

Taking into account the inequality $F'''(1) < \infty$ and using (3) and (12) to estimate $(\ln\varphi(u))'''_\varphi$ we obtain that for sufficiently small u and sufficiently large N, n as $n/N^2 \to 0$, $n/N \to \infty$

$$|(\ln\varphi(u))'''_u| \leqslant C_3(n/N)^5, \tag{13}$$

where the symbols $C_3, C_4, \ldots$ here and below denote positive constants. By Lemma 2.1, $\lambda \to 1$; therefore $B_\lambda \to 1$ and from (2), (1.9) and (2.1) we obtain $\sigma^2 \geqslant C_4(n/N)^3$. From this and (11), (13) it follows that $Q_1(u) \leqslant C_5\sqrt{n}/N$. If $0 < C_6 \leqslant n/N \leqslant C_7 < \infty$, then by analogy we get $Q_1(u) \leqslant C_5/\sqrt{N} \to 0$. Let $n/N \to 0$, $N\lambda^j \to \infty$. As we saw above, $\lambda \to 0$, $m \to 0$; therefore $|1 - e^{iu}(F_\lambda(\varphi(u)))'_\varphi| \geqslant C_8 > 0$. From this and (3) it follows that $(\ln\varphi(u))'''_u = O(\lambda^j)$. Using this relation, (5), (11) and the inequality $\sigma^2 \geqslant C_5\lambda^j$, which follows from (1.3), (1.8), (2) we obtain $Q_1(u) \to 0$. Thus $Q_1(u) \to 0$ in all domains of variation of N and n. Therefore from (10) we get that the last term on the right-hand side of (9) tends to zero. The Lemma is proved.

Lemma 1 shows that the distribution of ν_N weakly converges to the normal law as $N, n \to \infty$, $n/N^2 \to 0$. We will prove now that in fact local convergence takes place. Since the parameter λ depends on N and sums $\nu_N = \nu^{(1)} + \cdots + \nu^{(N)}$ form an array scheme it follows that we cannot apply the Theorem 1.3.5 directly. That is why the necessary assertion will be proved below in Lemma 2. As above, let j denote the least natural number such that $p_j > 0$ and let l be a non-negative integer which is not divided by j and such that $p_{j+l} > 0$; if there is no such l we put $l = 0$.

Lemma 2. *Let $N, n \to \infty$ in such a way that $n/N^2 \to 0$, and let $\lambda = \lambda(N, n)$ be determined by the relation* (1.9), *$N\lambda^{j+l} \to \infty$, $F'''(1) < \infty$ if $n/N \to \infty$. Then for a non-negative h divisible by d*

$$\mathbf{P}\{\nu_N = N + h\} = \frac{d(1 + o(1))}{\sigma\sqrt{2\pi N}} \exp\left\{-\frac{(h-n)^2}{2\sigma^2 N}\right\}$$

uniformly in $(h - n)/(\sigma\sqrt{N})$ lying in any finite interval.

Proof. By the invertion formula we represent the probability $\mathbf{P}\{\nu_N = N + h\}$ as the integral

$$\mathbf{P}\{\nu_N = N + h\} = d(2\pi\sigma\sqrt{N})^{-1} \int\limits_{-d^{-1}\pi\sigma\sqrt{N}}^{d^{-1}\pi\sigma\sqrt{N}} e^{-izu}\varphi_N(u)du, \tag{14}$$

where $z = (h - n)/(\sigma\sqrt{N})$. Since

$$(2\pi)^{-1/2}e^{-z^2/2} = (2\pi)^{-1} \int\limits_{-\infty}^{\infty} \exp\{-izu - u^2/2\}du, \tag{15}$$

the difference

$$R = 2\pi[d^{-1}\sigma\sqrt{N}\,\mathbf{P}\{\nu_N = N + h\} - (2\pi)^{-1/2}e^{-z^2/2}]$$

can be rewritten as the sum of four integrals $R = I_1 + I_2 + I_3 + I_4$, where

$$\begin{aligned} I_1 &= \int\limits_{-A}^{A} e^{-izu}[\varphi_N(u) - e^{-u^2/2}]du, \\ I_2 &= \int\limits_{A<|u|\leqslant\varepsilon\sigma\sqrt{N}} e^{-izu}\varphi_N(u)du, \\ I_3 &= \int\limits_{\varepsilon\sigma\sqrt{N}\leqslant|u|\leqslant d^{-1}\pi\sigma\sqrt{N}} e^{-izu}\varphi_N(u)du, \\ I_4 &= -\int\limits_{A<|u|} \exp\{-izu - u^2/2\}du; \end{aligned} \tag{16}$$

the positive constants A and ε will be chosen later.

The Lemma will be proved if we show that by choosing sufficiently large N, n, the difference R can be made arbitrarily small. To this end we estimate the integrals I_1–I_4.

Under the hypotheses of Lemma 2, Lemma 1 is valid, hence $I_1 \to 0$. Furthermore,

$$|I_4| \leqslant \int\limits_{A<|u|} e^{-u^2/2} du; \tag{17}$$

therefore, I_4 can be made arbitrarily small by choosing a sufficiently large A.

Now we need to estimate the integrals I_2 and I_3. Below we consider them in the simplest case, namely, where $N, n \to \infty$ in such a way that $0 < C_1 \leqslant n/N \leqslant C_2 < \infty$, and then point out some features of the estimation of I_2 and I_3 in other situations.

Using (3), (7), (8) and (12) we see that the inequality $|Q(u)| \leqslant C_3$ is valid for sufficiently small u; hereafter $C_3, C_4, \ldots$ mean some positive constants. Using (2) we obtain $\sigma^3 \geqslant C_4$; then from (9)–(11) we get

$$|\varphi_N(u)| \leqslant e^{-C_5 u^2}$$

for $|u| < \varepsilon\sigma\sqrt{N}$ and sufficiently small ε. Therefore from (16) we derive the following estimate similar to (17)

$$|I_2| \leqslant \int\limits_{A<|u|} e^{-C_5 u^2} du. \tag{18}$$

Let us estimate the integral I_3. For $\varepsilon < |u| \leqslant \pi/d$, the inequality

$$|\varphi(u)| \leqslant e^{-C_6} \tag{19}$$

is valid, since the span of the distribution $\nu^{(1)}$ is equal to d, the function $\varphi(u)$ is continous in λ and, as seen from Lemma 2.1, $0 < C_7 \leqslant \lambda \leqslant C_8 < 1$. From (16) and (19) we obtain, using (2), that

$$|I_3| \leqslant 2\pi d^{-1}\sigma\sqrt{N}e^{-C_6 N} \leqslant C_9\sqrt{N}e^{-C_6 N} \to 0. \tag{20}$$

Thus, the Lemma is proved as $0 < C_1 \leqslant n/N \leqslant C_2 < \infty$. Now we turn to the consideration of the behaviour of I_2 and I_3 in other cases. To this end we divide the following proof into four parts.

1. The estimation of the integral I_3 as $N, n \to \infty$ in such a way that $n/N \to \infty$, $n/N^2 \to 0$. As above we easily deduce from Lemma 2.1 that the inequality (19) is valid. Hence we obtain the following relation which is analogous to (20)

$$|I_3| \leqslant C_9 N^2 e^{-C_6 N} \to 0. \tag{21}$$

2. The estimation of the integral I_3 as $N, n \to \infty$ in such a way that $n/N \to 0$. The characteristic function of the random variable $\nu^{(1)}$ is

$$\varphi(u) = \sum_{k=0}^{\infty} \mathbf{P}\{\nu^{(1)} = kd + 1\} \exp\{iu(kd+1)\}.$$

From this and (2.4) we obtain

$$\varphi(u) = \sum_{k=0}^{\infty} \frac{\lambda^{kd}}{F^{kd+1}(\lambda)} \mathbf{P}\{\nu_*^{(1)} = kd + 1\} \exp\{iu(kd+1)\}, \tag{22}$$

where $\nu_*^{(1)}$ is the number of vertices of the critical Galton–Watson branching process beginning with one particle for which the number of offspring of a particle has the distribution (1.1).

If $p_1 > 0$, then the inequality $p_0 > 0$ implies that $d = 1$. From Lemma 2.1 we obtain $\lambda \to 0$; therefore from (22) we find

$$\varphi(u) = F^{-1}(\lambda)\,\mathbf{P}\{\nu_*^{(1)} = 1\}e^{iu} + \lambda F^{-2}(\lambda)\,\mathbf{P}\{\nu_*^{(1)} = 2\}e^{2iu} + O(\lambda^2).$$

Using (1.2) we get

$$\varphi(u) = e^{iu}[1 - (p_1/p_0)\lambda(1 - e^{iu}) + O(\lambda^2)]$$

and

$$|\varphi(u)| = 1 - (p_1/p_0)(1 - \cos u)\lambda + O(\lambda^2).$$

Since $\varepsilon \leqslant |u/(\sigma\sqrt{N})| \leqslant \pi/d$ it follows from Lemma 2.1 that

$$|\varphi(u/(\sigma\sqrt{N}))| \leqslant \exp\{-C_{10}n/N\}. \tag{23}$$

As $j = 1$, using (2), (4), (6), (16), (23) and Lemma 2.1 we obtain

$$|I_3| \leqslant C_9\sqrt{n}e^{-C_{10}n} \to 0.$$

This relation can easily be generalized to the case $p_j > 0$ if the inequalities $p_{j+k} > 0$ are true only for natural k which are divided by j. Then $j = d$ and by (22)

$$\varphi(u) = F^{-1}(\lambda)\,\mathbf{P}\{\nu_*^{(1)} = 1\}e^{iu}$$
$$+\lambda^d F^{-(d+1)}(\lambda)\,\mathbf{P}\{\nu_*^{(1)} = d+1\}e^{iu(d+1)} + o(\lambda^d).$$

This equation, (1.2) and the relation $\lambda \to 0$ imply that

$$\varphi(u) = e^{iu}[1 - (p_d/p_0)\lambda^d(1 - e^{iud}) + o(\lambda^d)]$$

and

$$|\varphi(u)| = 1 - (p_d/p_0)\lambda^d(1 - \cos(ud)) + o(\lambda^d).$$

Since $\varepsilon < |u/(\sigma\sqrt{N})| \leqslant \pi/d$ it follows that

$$|\varphi(u/(\sigma\sqrt{N}))| \leqslant \exp\{-C_{11}\lambda^d\}; \tag{24}$$

therefore

$$|I_3| \leqslant 2d^{-1}\pi\sigma\sqrt{N}\exp\{-C_{11}\lambda^d N\}.$$

Using (2), (4) and the condition $N\lambda^d \to \infty$ we get

$$|I_3| \leqslant C_{12}\sqrt{\lambda^d N}\exp\{-C_{11}\lambda^d N\} \to 0. \tag{25}$$

Finally, let $p_1 = 0$, $p_j > 0$, $p_{j+l} > 0$. Again using (22) we obtain as above

$$\begin{aligned}|\varphi(u)| \leqslant \exp\{&-C_{13}\lambda^j(1 - \cos(uj)) \\ &- C_{14}\lambda^{j+l}(1 - \cos(u(j+l)))\}.\end{aligned} \tag{26}$$

We represent the integral I_3 as the sum of two integrals $I_3 = I_3' + I_3''$ where the integration domains of I_3' and I_3'' are defined as follows. We assume that $j = d(2L+1)$ for some natural L. Then the integration domain of I_3'' is constituted by the intervals of the form $(2kj^{-1}\sigma\sqrt{N}(\pi - \varepsilon), 2kj^{-1}\sigma\sqrt{N}(\pi + \varepsilon))$, where $k = 1, \ldots, L$ and similar intervals in the negative part of the variation u. Naturally, the integration domain of I_3' is the complement of the integration domain of I_3'' to the set of values u in I_3. If $j = 2dL$, then the integration domains of I_3' and I_3'' are constructed similarly, but the right-hand border of the integration domain of I_3'' is equal to $\pi\sigma\sqrt{N}/d$.

Using (26) to estimate I_3' we obtain

$$|\varphi(u)| < \exp\{-C_{15}\lambda^j(1-\cos(uj))\}. \tag{27}$$

Therefore the condition $\lambda^j N \to \infty$ implies that

$$|I_3'| \leqslant C_6\sqrt{\lambda^j N}\exp\{-C_{17}\lambda^j N\} \to 0. \tag{28}$$

Now estimate I_3''. We consider the part of this integral $I_3''(1)$ which corresponds to the first interval in the positive argument domain, i.e.

$$I_3''(1) = \int\limits_{2j^{-1}(\pi-\varepsilon)\sigma\sqrt{N}}^{2j^{-1}(\pi+\varepsilon)\sigma\sqrt{N}} e^{-izu}\varphi_N(u)du.$$

Since

$$|I_3''(1)| \leqslant \int\limits_{2j^{-1}(\pi-\varepsilon)\sigma\sqrt{N}}^{2j^{-1}(\pi+\varepsilon)\sigma\sqrt{N}} |\varphi(u/(\sigma\sqrt{N}))|^N du,$$

it follows that

$$|I_3''(1)| \leqslant C_{18}\sigma\sqrt{N}\int\limits_{-2\varepsilon/j}^{2\varepsilon/j} |\varphi(y+2\pi/j)|^N dy. \tag{29}$$

Expanding $1-\cos(uj)$ into a series in the neighbourhood of $u = 2\pi/j$ we obtain

$$1-\cos(uj) = j^2(u-2\pi/j)^2 + O((u-2\pi/j)^3)$$

as $u \to 2\pi/j$. Therefore for sufficiently small ε and sufficiently large N, n

$$\exp\{-C_{13}\lambda^j(1-\cos(uj))\} \leqslant \exp\{-C_{19}\lambda^j(u-2\pi/j)^2\}.$$

From this inequality and (26) we get

$$|\varphi(u)| \leqslant \exp\{-C_{20}\lambda^{j+l} - C_{19}\lambda^j(u-2\pi/j)^2\}. \tag{30}$$

This relation together with (28) and (4) implies that

$$|I_3''(1)| \leqslant C_{21}\sigma\sqrt{N}\exp\left\{-C_{20}\lambda^{j+l}N\right\}\int\limits_{-2\varepsilon/j}^{2\varepsilon/j}\exp\left\{-C_{19}\lambda^j N y^2\right\}dy$$

$$\leqslant C_{22}\sqrt{\lambda^j N}\exp\left\{-C_{20}\lambda^{j+l}N\right\}\int\limits_{-2\varepsilon/j}^{2\varepsilon/j}\exp\left\{-C_{19}\left(y\sqrt{\lambda^j N}\right)^2\right\}dy$$

$$\leqslant C_{23}\exp\left\{-C_{20}\lambda^{j+l}N\right\}\int\limits_{-\infty}^{\infty}\exp\{-C_{24}y^2\}dy.$$

From this and the condition $N\lambda^{j+l} \to \infty$ it follows that $I_3''(1) \to 0$. Let us consider by analogy

$$I_3''(2) = \int\limits_{4j^{-1}(\pi-\varepsilon)\sigma\sqrt{N}}^{4j^{-1}(\pi+\varepsilon)\sigma\sqrt{N}} e^{-izu}\varphi_N(u)du.$$

Hence

$$|I_3''(2)| \leqslant \int\limits_{4j^{-1}(\pi-\varepsilon)\sigma\sqrt{N}}^{4j^{-1}(\pi+\varepsilon)\sigma\sqrt{N}} |\varphi(u/(\sigma\sqrt{N}))|^N du$$

$$= \sigma\sqrt{N} \int\limits_{-4\varepsilon/j}^{4\varepsilon/j} |\varphi(y + 4\pi/j)|^N dy.$$

Therefore, using the expansion

$$1 - \cos(uj) = j^2(u - 4\pi/j)^2 + O((u - 4\pi/j)^3)$$

we obtain

$$|I_3''(2)| \leqslant C_{23} \exp\{-C_{20}\lambda^{j+l}N\} \int\limits_{-\infty}^{\infty} \exp\{-C_{24}y^2\}dy \to 0.$$

It is easy to see that similar estimates can be obtained for parts $I_3''(3), \ldots, I_3''(L)$ and for parts of I_3'' with the negative integration domains of the variation u. That is why

$$|I_3''| \leqslant 2(|I_3''(1)| + \cdots + |I_3''(L)|) \to 0. \tag{31}$$

3. The estimation of the integral I_2 for $N, n \to \infty$ in such a way that $n/N \to 0$. By analogy with the proof of Lemma 1 we see that $|1 - e^{iu}(F_\lambda(\varphi(u)))'_\varphi| \geqslant C_{25} > 0$ and from (3) we get $(\ln \varphi(u))'''_u = O(\lambda^j)$. Furthermore, from (2) and (4) we obtain $\sigma^2 \geqslant C_{26}\lambda^j$ and (6)–(11) imply that

$$|\varphi_N(u)| \leqslant \exp\{-C_{27}u^2\}. \tag{32}$$

This relation means that the estimate (18) is valid for the integral I_2.

4. The estimation of the integral I_2 for $N, n \to \infty$ in such a way that $n/N \to \infty$, $n/N^2 \to 0$. We represent I_2 as the sum $I_2 = I_2' + I_2''$ where the integration domains of the integrals I_2' and I_2'' are the sets $\{u:\ A < |u| \leqslant \varepsilon_1 N/\sqrt{n}\}$, $\{u:\ \varepsilon_1 N/\sqrt{n} < |u| \leqslant \varepsilon\sigma\sqrt{N}\}$, the positive constant ε_1 will be chosen later. We begin with the estimation of I_2'. From (7) and (8) we obtain that, for sufficiently small ε_1 and $|u/(\sigma\sqrt{N})| < \varepsilon_1$,

$$\ln\varphi\left(\frac{u}{\sigma\sqrt{N}}\right) = \frac{iua}{\sigma\sqrt{N}} - \frac{u^2}{2N} + \frac{u^3}{6\sigma^3 N^{3/2}} Q\left(\frac{u}{\sigma\sqrt{N}}\right), \tag{33}$$

where

$$|Q(u/(\sigma\sqrt{N}))| \leqslant 2 \max_{|\tau| \leqslant |u/(\sigma\sqrt{N})|} |(\ln\varphi(\tau))'''_\tau|. \tag{34}$$

From Lemma 2.1 and (2), (1.9) it follows that $\sigma^2 \geqslant C_{28}(n/N)^3$; hence $u/(\sigma\sqrt{N}) \to 0$ in the integration domain of I_2'. From this and (13) we obtain

$$\left|\frac{u}{\sigma^3 N^{3/2}} Q\left(\frac{u}{\sigma\sqrt{N}}\right)\right| \leqslant \varepsilon_1 C_{29}/N. \tag{35}$$

Therefore (6), (33) and (34) imply that for sufficiently small ε_1

$$|\varphi_N(u)| \leqslant \exp\{-C_{30}u^2\}$$

and

$$|I_2'| \leqslant 2\int_A^\infty \exp\{-C_{30}u^2\}du. \tag{36}$$

This relation can usually be made arbitrarily small by choosing a sufficiently large A.

Now we estimate I_2''. By Lemma 1.3.3

$$\varphi(u/(\sigma\sqrt{N})) = 1 - \sqrt{2(1-\exp\{iu/(\sigma\sqrt{N})\})/B_\lambda}(1+o(1));$$

therefore, as $|u/(\sigma\sqrt{N})| < \varepsilon$

$$|\varphi(u/(\sigma\sqrt{N}))| \leqslant \exp\{-C_{31}\sqrt{|u|/(\sigma\sqrt{N})}\} \tag{37}$$

and from (1.9), (2.1) and (2) we obtain

$$|I_2''| \leqslant 2\int_{\varepsilon_1 N/\sqrt{n}}^\infty \exp\{-C_{31}\sqrt{u}(N^2/n)^{3/4}\}du. \tag{38}$$

The last expression tends to zero because $N/\sqrt{n} \to \infty$. Lemma 2 is proved.

Lemma 2 describes the limit behaviour of ν_N as $N, n \to \infty$, $n/N^2 \to 0$ in the subcritical case $\lambda < 1$. Let us consider the asymptotics of ν_N as $\lambda = 1$. We denote as B the variance of the distribution (1.3) in this case.

Lemma 3. *Let $N \to \infty$, $z = (N+h)BN^{-2}$, where h is a natural number divisible by d, $\lambda = 1$. Then*

$$B^{-1}N^2\,\mathbf{P}\{B\nu_N/N^2 = z\} = d(2\pi)^{-1/2}z^{-3/2}e^{-1/(2z)}(1+o(1))$$

uniformly in z, $0 < z_0 \leqslant z \leqslant z_1 < \infty$.

Proof. Since the characteristic function of the random variable ν_N is $\varphi^N(u)$ from Lemma 1.3.3 we see that as $N, n \to \infty$ for any fixed u

$$\varphi^N(uB/N^2) \to \exp\{-\sqrt{-2iu}\} = \exp\{-\sqrt{|u|}(1 - iu/|u|)\} \tag{39}$$

and the last expression is the characteristic function of the stable law with the exponent $\alpha = 1/2$. The maximum span of the distribution of $\nu_N = \nu^{(1)} + \cdots + \nu^{(N)}$ is d; hence from (39) and Theorem 1.3.6 we obtain that for this sum the local limit theorem on convergence to stable distribution with $\alpha = 1/2$ and density $(2\pi x^3\exp\{1/x\})^{-1/2}$ is valid. The Lemma is proved.

Lemma 4. *Let $N \to \infty$, $z = B(N+h)N^{-2} \to \infty$, where h is a natural number divisible by d, $\lambda = 1$. Then*

$$B^{-1}N^2\,\mathbf{P}\{B\nu_N/N^2 = z\} = d(2\pi z^3)^{-1/2}(1+o(1)).$$

Proof. By virtue of Lemma 1.3.12 as $k \to \infty$

$$\mathbf{P}\{\nu^{(1)} = kd+1\} = O(k^{-3/2}).$$

From this relation and (39) we get that the sum $\nu_N = \nu^{(1)} + \cdots + \nu^{(N)}$ satisfies the hypothesis of Theorem 1.3.7 which implies the validity of Lemma 4 as $d = 1$. If $d \neq 1$ then the assertion of Lemma 4 is valid too (see the proofs of Lemmas 5.2 and 3.3.2).

4. The convergence of the sum of auxiliary random variables to the normal law

In this Section we will consider the limit behaviour of the probability $\mathbf{P}\{\nu_N^{(r)} = N+n\}$ from the right-hand side of the relation (1.8) as $N, n \to \infty$ in such a way that $n/N^2 \to 0$. By (1.6) the random variable $\nu_{(N)}^{(r)}$ is closely connected with the random variable ν_N the limit distributions of which are obtained in Section 3. The results on $\nu_{(N)}^{(r)}$ will be proved below by analogy with the results on ν_N taking into consideration the behaviour of the parameter r.

Let $\varphi_r(u)$ be the characteristic function of the random variable $(\nu_N^{(r)} - N - n)/(\sigma\sqrt{N})$.

Lemma 1. *Suppose $N, n \to \infty$, $n/N^2 \to 0$, $\lambda = \lambda(N,n)$ is determined by the relation (1.9), $F'''(1) < \infty$ if $n/N \to \infty$ and let $r = r(N,n)$ take values which are divided by d and vary in such a way that $NP_r \to \gamma$, where γ is a positive constant and P_r is defined in Section 1. Then*

$$\varphi_r(u) \to e^{-u^2/2}$$

uniformly in u lying in any finite interval. This assertion remains true for $\gamma = 0$ if $n/N \to 0$, $NP_{r-d} \geqslant C > 0$, $N\lambda^j \to \infty$, where j is the least natural number such that $p_j > 0$.

Proof. It is easy to see that

$$\begin{aligned} \mathbf{E}\exp\{iu\nu_{[r]}^{(1)}\} = (1-P_r)^{-1}(\varphi(u) \\ -\sum_{k=1}^{\infty}\mathbf{P}\{\nu^{(1)} = r+kd+1\}\exp\{iu(r+kd+1)\}; \end{aligned} \tag{1}$$

therefore

$$\begin{aligned} \varphi_r(u) = \exp\left\{-\frac{i(N+n)u}{\sigma\sqrt{N}}\right\}(1-P_r)^N \\ \times\left(\varphi\left(\frac{u}{\sigma\sqrt{N}}\right) - \sum_{k=1}^{\infty}\mathbf{P}\{\nu^{(1)} = r+kd+1\}\exp\left\{\frac{iu(r+kd+1)}{\sigma\sqrt{N}}\right\}\right)^N. \end{aligned} \tag{2}$$

Using (3.5)–(3.8) it can be proved similarly to the way Lemma 3.1 was proved that

$$\exp\{-iua/(\sigma\sqrt{N})\}\varphi(u/(\sigma\sqrt{N})) = 1 - u^2/(2N) + o(1/N). \tag{3}$$

Hence

$$\varphi_r(u) = e^{-u^2/2}(1-P_r)^{-N}(1+o(1))$$
$$\times \left(1-(1+o(1))\sum_{k=1}^{\infty}\mathbf{P}\{\nu^{(1)}=r+kd+1\}\exp\left\{\frac{iu(r+kd+1)}{\sigma\sqrt{N}}\right\}\right)^N. \tag{4}$$

From the equation

$$\int_0^x e^{it}dt = i^{-1}(e^{ix}-1)$$

it follows that $|e^{ix}-1| \leqslant |x|$; therefore

$$\sum_{k=1}^{\infty}\mathbf{P}\{\nu^{(1)}=r+kd+1\}\exp\{iu(r+kd+1)/(\sigma\sqrt{N})\} = P_r + Q(u), \tag{5}$$

where

$$|Q(u)| \leqslant |u|(\sigma\sqrt{N})^{-1}\sum_{k=1}^{\infty}(r+kd+1)\,\mathbf{P}\{\nu^{(1)}=r+kd+1\}. \tag{6}$$

From (4)–(6) we see that to prove the Lemma it suffices to get that

$$(\sigma\sqrt{N})^{-1}\sum_{k=1}^{\infty}(r+kd+1)\,\mathbf{P}\{\nu^{(1)}=r+kd+1\} = o(N^{-1}). \tag{7}$$

Further, we consider three cases: $n/N \to 0$, $n/N \to b$, where b is some positive constant and $n/N \to \infty$, $n/N^2 \to 0$.

Firstly let $n/N \to 0$. From Lemma 2.1 it follows that $\lambda \to 0$ and using (2.4) we obtain

$$\frac{1}{\sigma\sqrt{N}}\sum_{k=1}^{\infty}(r+kd+1)\,\mathbf{P}\{\nu^{(1)}=r+kd+1\}$$
$$= O\left(\frac{r+s+1}{\sigma\sqrt{N}}\,\mathbf{P}\{\nu^{(1)}=r+s+1\}\right), \tag{8}$$

where s is the least natural number which is divided by d and such that $\mathbf{P}\{\nu^{(1)} = r+s+1\} > 0$. It is clear that if such s does not exist then the last term in the right-hand side of (5) is equal to zero and the Lemma is true. From Lemma 1.3.12 we see that if $r \to \infty$ then $s = d$ and if r is fixed then s is fixed too. The condition $NP_r \to \gamma$ and (2.7) imply that $N\,\mathbf{P}\{\nu^{(1)} = r+s+1\} \to \gamma$. From this, (8), (3.2), (3.4) we get

$$(\sigma\sqrt{N})^{-1}\sum_{k=1}^{\infty}(r+kd+1)\,\mathbf{P}\{\nu^{(1)}=r+kd+1\} \leqslant C_1 r/(N\sqrt{N\lambda^j}); \tag{9}$$

here and below $C_1, C_2, \ldots$ are some positive constants. Taking into account the condition $N\lambda^j \to \infty$ from (9) we obtain that (7) is valid for any fixed r. Let $r \to \infty$.

We claim that under the conditions of the Lemma, r does not tend to infinity too quickly. Indeed,

$$P_r = \sum_{k=1}^{\infty} \mathbf{P}\{\nu^{(1)} = r + kd + 1\}$$
$$= \mathbf{P}\{\nu^{(1)} = r + d + 1\} \left(1 + \sum_{k=2}^{\infty} \frac{\mathbf{P}\{\nu^{(1)} = r + kd + 1\}}{\mathbf{P}\{\nu^{(1)} = r + d + 1\}}\right) \tag{10}$$

and from (2.4) and the relation $\lambda \to 0$ we obtain

$$P_r = \lambda^{r+d} F^{-(r+d+1)}(\lambda)\, \mathbf{P}\{\nu_*^{(1)} = r + d + 1\}(1 + o(1)).$$

From this we see that if $NP_r \to \gamma$ then $NP_{r-d} \to \infty$ and if $NP_r \to 0$ then by the condition of the Lemma, $NP_{r-d} \geqslant C > 0$. From Lemmas 1.3.12 and 2.1 it is not hard to get

$$N\,\mathbf{P}\{\nu^{(1)} = r + 1\} \leqslant C_2 N r^{-3/2} \lambda^{C_3 r}. \tag{11}$$

Therefore $r = o(\sqrt{N\lambda^j})$ as otherwise (10), (11) and (2.4) imply that $N\,\mathbf{P}\{\nu^{(1)} = r + 1\} \to 0$ and $NP_{r-d} \to 0$ which is impossible. Thus from (9) we obtain that (7) is true as $r \to \infty$.

Let $n/N \to b$. By corollary 2.1

$$N\,\mathbf{P}\{\nu^{(1)} = r + 1\} \to \gamma\alpha^{-1}(1 - \alpha), \tag{12}$$

where α is defined by (1.10). From (2.4) we see that in this case $r \to \infty$. Since

$$(\sigma\sqrt{N})^{-1} \sum_{k=1}^{\infty} (r + kd + 1)\, \mathbf{P}\{\nu^{(1)} = r + kd + 1\}$$
$$= \frac{(r+1)\,\mathbf{P}\{\nu^{(1)} = r + 1\}}{\sigma\sqrt{N}} \sum_{k=1}^{\infty} \frac{(r + kd + 1)\,\mathbf{P}\{\nu^{(1)} = r + kd + 1\}}{(r+1)\,\mathbf{P}\{\nu^{(1)} = r + 1\}}$$

and by (2.11)

$$\frac{(r + kd + 1)\,\mathbf{P}\{\nu^{(1)} = r + kd + 1\}}{(r+1)\,\mathbf{P}\{\nu^{(1)} = r + 1\}} \to \alpha^k,$$

we obtain, using (1.9), (2.1), (3.2) and (12), that

$$(\sigma\sqrt{N})^{-1} \sum_{k=1}^{\infty} (r + kd + 1)\, \mathbf{P}\{\nu^{(1)} = r + kd + 1\}$$
$$= (r\gamma/\sqrt{B_{\lambda_b}})((1 + b)N)^{-3/2}(1 + o(1)), \tag{13}$$

where λ_b is the solution of the equation (1.9). From Lemma 1.3.12 and (2.4) we get

$$\mathbf{P}\{\nu^{(1)} = r + 1\} = \lambda^r d(F^{r+1}(\lambda) r^{3/2} \sqrt{2\pi B})^{-1}(1 + o(1)).$$

Therefore (12) implies that

$$\frac{Nd}{F(\lambda) r^{3/2} \sqrt{2\pi B}} \left(\frac{\lambda}{F(\lambda)}\right)^r \to \gamma\alpha^{-1}(1 - \alpha).$$

Taking the logarithm of this relation and dividing both sides of the resulting relation by r we see that $(\ln N)/r \to C_4 > 0$. This means that $r = O(\ln N)$ and from (13) we obtain (7).

Finally, let $n/N \to \infty$, $n/N^2 \to 0$. By (1.9)

$$\lambda = (n/(N+n))(1+o(1)) \tag{14}$$

and from (14), (2.19) and the condition $n/N^2 \to 0$ we obtain $N\sqrt{\beta} \to \infty$. Using (2.18) and the condition $NP_r \to \gamma$ we get

$$r\beta \to \infty. \tag{15}$$

By analogy with the proof of the equation (2.15) we can obtain that

$$\sum_{k=1}^{\infty}(r+kd+1)\,\mathbf{P}\{\nu^{(1)} = r+kd+1\}$$
$$= d(2\pi B\beta)^{-1/2}(1+o(1))\sum_{k=1}^{\infty}(r+kd)^{-1/2}\exp\{-(r+kd)\beta\}.$$

Using this relation and Lemma 2.4 we find the estimation similar to (2.18):

$$\sum_{k=1}^{\infty}(r+kd+1)\,\mathbf{P}\{\nu^{(1)} = r+kd+1\}$$
$$= (1+o(1))(2\pi B\beta)^{-1/2}\int_{r\beta}^{\infty} y^{-1/2}e^{-y}dy.$$

This relation, (15) and Lemma 2.4 imply that

$$\sum_{k=1}^{\infty}(r+kd+1)\,\mathbf{P}\{\nu^{(1)} = r+kd+1\} = (2\pi B\beta^2 r)^{-1/2}e^{-r\beta}(1+o(1)).$$

Hence from (1.9), (2.1) and (3.2) we get

$$\begin{aligned}(\sigma\sqrt{N})^{-1}&\sum_{k=1}^{\infty}(r+kd+1)\,\mathbf{P}\{\nu^{(1)} = r+kd+1\}\\ &= N(e^{r\beta}B\beta n^{3/2}\sqrt{2\pi r})^{-1}(1+o(1)).\end{aligned} \tag{16}$$

From (2.18) and Lemma 2.4 we obtain the following relation:

$$P_r = (e^{r\beta}r^{3/2}\beta\sqrt{2\pi B})^{-1}(1+o(1)). \tag{17}$$

Since $NP_r \to \gamma$ it follows as $w = r\beta$ that

$$\frac{N\beta^{1/2}}{w^{3/2}\sqrt{2\pi B}}e^{-w} = \gamma + o(1). \tag{18}$$

The application of (15) yields $w^{1/2}e^{-w} \to 0$; hence, combining (14), (18) and (2.19) we obtain

$$N^2/(nw^2) \to \infty. \tag{19}$$

Now if we recall (2.19), (16) and (17) we get

$$(\sigma\sqrt{N}P_r)^{-1}\sum_{k=1}^{\infty}(r+kd+1)\,\mathbf{P}\{\nu^{(1)}=r+kd+1\}$$
$$=2w\sqrt{n}(B^{3/2}N)^{-1}(1+o(1)).$$

This equality and (19) imply that

$$(\sigma\sqrt{N})^{-1}\sum_{k=1}^{\infty}(r+kd+1)\,\mathbf{P}\{\nu^{(1)}=r+kd+1\}=o(P_r).$$

Now by the relation $P_r = O(N^{-1})$ we obtain (7). This completes the proof of Lemma 1.

By Lemma 1 the distributions of the random variable $\nu_N^{(r)}$ weakly converge to the normal law. Now we will prove the local convergence of these distributions.

Lemma 2. *Let the hypotheses of Lemma 1 be satisfied. Let $N\lambda^{j+l}\to\infty$, where l is the least natural number which is not divided by j and such that $p_{j+l}>0$; if there is no such l, we put $l=0$. Then for non-negative integers h divisible by d*

$$\mathbf{P}\{\nu_N^{(r)}=N+h\}=\frac{d(1+o(1))}{\sigma\sqrt{2\pi N}}\exp\left\{-\frac{(h-n)^2}{2\sigma^2N}\right\}$$

uniformly in $(h-n)/(\sigma\sqrt{N})$ lying in any finite interval.

Proof. We follow the scheme for proving Lemma 3.2. We represent the probability $\mathbf{P}\{\nu_N^{(r)}=N+h\}$ as the integral

$$\mathbf{P}\{\nu_N^{(r)}=N+h\}=\frac{d}{2\pi\sigma\sqrt{N}}\int_{-d^{-1}\pi\sigma\sqrt{N}}^{d^{-1}\pi\sigma\sqrt{N}}e^{-izu}\varphi_r(u)du,$$

where $z=(h-n)/(\sigma\sqrt{N})$. Using (3.15) the difference

$$R=2\pi[d^{-1}\sigma\sqrt{N}\,\mathbf{P}\{\nu_N^{(r)}=N+h\}-(2\pi)^{-1/2}e^{-u^2/2}]$$

can be rewritten as the sum of four integrals $R=I_1+I_2+I_3+I_4$, where

$$\begin{aligned}
I_1&=\int_{-A}^{A}e^{-izu}[\varphi_r(u)-e^{-u^2/2}]du,\\
I_2&=\int_{A<|u|\leqslant\varepsilon\sigma\sqrt{N}}e^{-izu}\varphi_r(u)du,\\
I_3&=\int_{\varepsilon\sigma\sqrt{N}<|u|\leqslant d^{-1}\pi\sigma\sqrt{N}}e^{-izu}\varphi_r(u)du,\\
I_4&=-\int_{A<|u|}\exp\{-izu-u^2/2\}du;
\end{aligned}\qquad(20)$$

the positive constants A and ε will be chosen later. By Lemma 1, $I_1 \to 0$ and for I_4 we can use the estimation (3.17).

To estimate the integrals I_2 and I_3, we firstly consider the case $n/N \leqslant C_1 < \infty$. By analogy with the proof of (3.18) and (3.30) it is not hard to find that in the integration domain of I_2

$$\varphi(u/(\sigma\sqrt{N})) \leqslant \exp\{-C_2u^2/N\}; \tag{21}$$

hereafter $C_2, C_3, \dots$ mean some positive constants. It is clear that for sufficiently large N, n

$$\left|\sum_{k=1}^{\infty} \mathbf{P}\{\nu^{(1)} = r + kd + 1\} \exp\{iu(r + kd + 1)/(\sigma\sqrt{N})\}\right| \leqslant P_r \leqslant C_3/N. \tag{22}$$

From this and (2), (21) we get

$$|\varphi_r(u)| \leqslant (1 - P_r)^{-N} \left(\exp\{-C_2u^2/N\} + C_3/N\right)^N.$$

Therefore, using relation $NP_r \to \gamma$ we obtain

$$\begin{aligned} |I_2| &\leqslant C_5 \int_A^{\varepsilon\sigma\sqrt{N}} (\exp\{-C_2u^2/N\} + C_3/N)^N du \\ &\leqslant C_5(1 + C_3N^{-1}e^{C_6})^N \int_A^{\varepsilon\sigma\sqrt{N}} e^{-C_2u^2} du \leqslant C_7 \int_A^{\infty} e^{-C_2u^2} du. \end{aligned} \tag{23}$$

It is clear that I_2 can be made arbitrarily small by choosing a sufficiently large A.

Let $\varepsilon < |u| \leqslant \pi/d$. If $n/N \geqslant C_1 > 0$, then the inequality (3.19) is valid; therefore from (2), (22) and the condition $NP_r \to \gamma$ we find that

$$|\varphi_r(u)| \leqslant (1 - P_r)^N (e^{-C_8} + C_3/N)^N \leqslant C_9 e^{-C_8 N}$$

and

$$|I_3| \leqslant C_{10}\sigma\sqrt{N}e^{-C_8N} \leqslant C_{11}\sqrt{N}e^{-C_8N} \to 0.$$

If $n/N \to 0$ then by analogy with the proof of Lemma 3.2 we will consider two cases. If $j = 1$ or $p_1 = 0$ and all natural k such that $p_{j+k} > 0$ are divided by j, then $j = d$ and from (3.23), (3.24), (2) and (22) we obtain that

$$|\varphi_r(u)| \leqslant (1 - P_r)^N (\exp\{-C_{12}\lambda^j\} + C_3/N)^N \leqslant C_{13} \exp\{-C_{12}\lambda^j N\}.$$

This relation and (3.2), (3.4) imply that

$$|I_3| \leqslant C_{14}\sigma\sqrt{N} \exp\{-C_{12}\lambda^j N\} \leqslant C_{14}\sqrt{\lambda^j N} \exp\{-C_{12}\lambda^j N\} \to 0.$$

Let $p_1 = 0$, $p_j > 0$, $l > 0$. We divide the integral I_3 into the sum of two integrals $I_3 = I_3' + I_3''$, where the integration domains of the integrals I_3' and I_3'' coincide with the domains of similar integrals in the proof of Lemma 3.2. From (3.27), (2) and (22) we obtain that in the integration domain of I_3'

$$|\varphi_r(u)| \leqslant (1 - P_r)^N (\exp\{-C_{15}\lambda^j\} + C_3/N)^N \leqslant C_{16} \exp\{-C_{15}\lambda^j N\}.$$

From this we see by analogy with (3.28) that

$$|I_3'| \leqslant C_{17}\sqrt{\lambda^j N} \exp\{-C_{15}\lambda^j N\} \to 0.$$

Now estimate I_3''. We consider the integral $I_3''(1)$ which is determined in the proof of Lemma 3.2. Using (2) and (22) we obtain

$$|I_3''(1)| \leqslant C_{18}(1-P_r)^N \int\limits_{2j^{-1}(\pi-\varepsilon)\sigma\sqrt{N}}^{2j^{-1}(\pi+\varepsilon)\sigma\sqrt{N}} \left(\varphi\left(\frac{u}{\sigma\sqrt{N}}\right)+\frac{C_3}{N}\right)^N du.$$

From (3.22) and the relation $\lambda \to 0$ we get $|\varphi(u/(\sigma\sqrt{N}))| \geqslant C_{19} > 0$; therefore

$$|\varphi(u/(\sigma\sqrt{N}))+C_3/N| \leqslant |\varphi(u/(\sigma\sqrt{N}))|(1+C_3/(C_{19}N))$$

and

$$|I_3''(1)| \leqslant C_{20} \int\limits_{2j^{-1}(\pi-\varepsilon)\sigma\sqrt{N}}^{2j^{-1}(\pi+\varepsilon)\sigma\sqrt{N}} \left|\varphi\left(\frac{u}{\sigma\sqrt{N}}\right)\right|^N du.$$

The last relation implies (3.29) and, as in the proof of Lemma 3.2, we get (3.30) and (3.31).

To conclude the proof it remains to consider the integrals I_2 and I_3 as $n/N \to \infty$, $n/N^2 \to 0$. We present the integration domains of I_2 as the union of two parts:

$$S_1 = \{u:\ A < |u| \leqslant \varepsilon_1 N/\sqrt{n}\}, \quad S_2 = \{u:\ \varepsilon_1 N/\sqrt{n} < |u| \leqslant \varepsilon\sigma\sqrt{N}\}.$$

For sufficiently small ε_1 the inequality (3.35) is valid in S_1. From this and (3.33), (3.34) we obtain

$$|\varphi(u/(\sigma\sqrt{N}))| \leqslant \exp\{-C_{21}u^2/N\}.$$

Therefore, from (2), (22) we get

$$\left|\int\limits_{S_1} e^{-izu}\varphi_r(u)du\right| \leqslant C_{22}\int\limits_A^{\infty} e^{-C_{23}u^2}du. \tag{24}$$

The estimation (3.37) is true in S_2 and from (2), (4) we see that

$$\left|\int\limits_{S_2} e^{-izu}\varphi_r(u)du\right| \leqslant C_{24}\int\limits_{\varepsilon_1 N\sqrt{n}}^{\infty} \exp\{-C_{25}\sqrt{u}(N^2/n)^{3/4}\}du \to 0.$$

From this and (24) we have that for sufficiently small ε_1 and sufficiently large N, n, A the integral I_2 can be made arbitrarily small.

We use (3.19) to estimate the integral I_3. Then from (4), (20) and (22) we obtain by analogy with (3.20) that

$$|I_3| \leqslant C_{25}\sqrt{N}e^{-C_{26}N} \to 0.$$

This completes the proof of Lemma 2.

5. The asymptotics of the sum of auxiliary random variables in the critical case

In the preceding Section we proved the local convergence of $\nu_N^{(r)}$ to the normal law as $N, n \to \infty$ such that $n/N^2 \to 0$. Now we will study the limit behaviour of this random variable in the domain $n/N^2 \geqslant C > 0$ for the critical process G.

Lemma 1. *Let $N, n \to \infty$ in such a way that $Bn/N^2 \to \gamma$, where γ is a positive constant. Let $v = (N+h)B/N^2$, where h is a natural number, divisible by d. If r takes values which are divided by d and $r = zn + O(1)$, z is a fixed positive number, then*

$$N^2 B^{-1} \mathbf{P}\{N^{-2} B \nu_N^{(r)} = v\} \to \frac{d}{\sqrt{2\pi}} e^{E(0,z)} \sum_{k=0}^{\infty} \frac{(-1)^k}{k!} I_k(\gamma z, v)$$

uniformly in v, $0 < v_0 \leqslant v \leqslant v_1$ with any finite v_0, v_1, where

$$E(u,z) = (2\pi)^{-1/2} \int_{\gamma z}^{\infty} y^{-3/2} e^{iuy} dy \tag{1}$$

and integrals $I_k(u,v)$, $k = 0, 1, \ldots$ are determined in Theorem 1.4.

Proof. Let $\Psi_r(u)$ be the characteristic function of the random variable $B\nu_N^{(r)}/N^2$. It is easy to see by analogy with (4.2) that

$$\Psi_r(u) = (1 - P_r)^{-N}$$
$$\times \left(\varphi(uBN^{-2}) - \sum_{k=1}^{\infty} \mathbf{P}\{\nu^{(1)} = r + kd + 1\} \exp\{iuB(r+kd+1)N^{-2}\} \right)^N .$$

Therefore, from (3.39) we obtain for any fixed u

$$\Psi_r(u) = (1 - P_r)^{-N} \exp\{-\sqrt{-2iu}\}$$
$$\times \left[1 - (1+o(1)) \sum_{k=1}^{\infty} \mathbf{P}\{\nu^{(1)} = r + kd + 1\} \exp\{iuB(r+kd+1)N^{-2}\} \right]^N . \tag{2}$$

Using Lemma 1.3.12 it is not hard to get

$$\sum_{k=1}^{\infty} \mathbf{P}\{\nu^{(1)} = r + kd + 1\} \exp\{iuB(r+kd+1)N^{-2}\}$$
$$= d(2\pi B)^{-1/2}(1+o(1)) \sum_{k=1}^{\infty} (r+kd+1)^{-3/2}$$
$$\times \exp\{iuB(r+kd+1)N^{-2}\} = (N\sqrt{2\pi B})^{-1}(1+o(1)) \tag{3}$$
$$\times \int_{\gamma z/B}^{\infty} y^{-3/2} \exp\{iuBy\} dy = N^{-1} E(u,z)(1+o(1)),$$

where the function $E(u,z)$ is determined by (1).

Similary we can find using (2.6) that

$$P_r = N^{-1} E(0,z)(1+o(1)). \tag{4}$$

From (2), (3), (4) and (3.39) it follows that

$$\Psi_r(u) = \exp\{-\sqrt{-2iu} - E(u,z) + E(0,z)\} + o(1) \tag{5}$$

uniformly in u lying in any finite interval. Since $E(u, z)$ is the Fourier transform of

$$f(t) = \begin{cases} (\sqrt{2\pi t^3})^{-1} & \text{for } t \geqslant \gamma z, \\ 0 & \text{for } t < \gamma z, \end{cases}$$

expanding $\exp\{-E(u, z)\}$ in powers of $E(u, z)$ and keeping in mind that $\exp\{-\sqrt{-2iu}\}$ is the characteristic function of the distribution with the density $(2\pi)^{-1/2} x^{-3/2} e^{-1/(2x)}$, we derive from Theorem 1.3.6 that $\Psi_r(u)$ converges to the characteristic function of the distribution with the density

$$g(v) = \frac{1}{\sqrt{2\pi}} e^{E(0,z)} \sum_{k=0}^{\infty} \frac{(-1)^k}{k!} I_k(\gamma z, v). \tag{6}$$

The parameter r depends on N and n; therefore the sum $\nu_N^{(r)} = \nu_{[r]}^{(1)} + \cdots + \nu_{[r]}^{(N)}$ forms an array scheme. That is why in contrast to the proof of Lemma 3.3 we cannot use the Theorem 1.3.6 to prove the local convergence of the distribution $\nu_N^{(r)}$. Further, we follow the well-known scheme of proving local limit theorems.

By the inversion formula, the probability $\mathbf{P}\{\nu_N^{(r)} = N + h\} = \mathbf{P}\{N^{-2} B \nu_N^{(r)} = v\}$ can be represented in the form

$$\mathbf{P}\{N^{-2} B \nu_N^{(r)} = v\} = \frac{dB}{2\pi N^2} \int\limits_{-\pi N^2/(dB)}^{\pi N^2/(dB)} e^{-iuv} \Psi_r(u) du. \tag{7}$$

Using (5) we can obtain

$$|\exp\{-\sqrt{-2iu} - E(u, z) + E(0, z)\}| \leqslant C \exp\{-\sqrt{|u|}\}, \tag{8}$$

where C is some positive constant. From this we see that the function $\exp\{-\sqrt{-2iu}$ $-E(u, z) + E(0, z)\}$ is integrable in u. Using the inversion formula we find that

$$g(v) = (2\pi)^{-1} \int\limits_{-\infty}^{\infty} \exp\{-ivu - \sqrt{-2iu} - E(u, z) + E(0, z)\} du,$$

where the function $g(v)$ is determined by (6). From this and (7) we obtain

$$2\pi N^2 (dB)^{-1} \mathbf{P}\{N^{-2} B \nu_N^{(r)} = v\} - 2\pi g(v) = I_1 + I_2 + I_3 + I_4,$$

where

$$I_1 = \int_{-A}^{A} e^{-iuv}[\Psi_r(u) - \exp\{-\sqrt{-2iu} - E(u,z) + E(0,z)\}]du,$$

$$I_2 = \int_{A<|u|\leqslant \varepsilon N^2} e^{-iuv}\Psi_r(u)du,$$

$$I_3 = \int_{\varepsilon N^2<|u|\leqslant \pi N^2/(dB)} e^{-iuv}\Psi_r(u)du,$$

$$I_4 = -\int_{A<|u|} \exp\{-iuv - \sqrt{-2iu} - E(u,z) + E(0,z)\}du;$$

the positive constants A and ε will be chosen later.

The Lemma will be proved if we show that by choosing sufficiently large N, n, A and a sufficiently small ε each of these integrals can be made arbitrarily small.

From (5) it follows that for any fixed A the integral I_1 tends to zero. Using (8) we obtain

$$|I_4| \leqslant 2C\int_A^\infty e^{-\sqrt{u}}du$$

and the last expression can be made arbitrarily small by choosing a sufficiently large A.

By Lemma 1.3.3, $|\varphi(u)| \leqslant \exp\{-\sqrt{|u|/(2B)}\}$ as $|u| \leqslant \varepsilon$ for sufficiently small ε. Therefore, from (2) it follows that

$$|I_2| \leqslant 2(1-P_r)^{-N}\int_A^{\varepsilon N^2}(\exp\{-\sqrt{u/(2N^2)}\} + P_r)^N du$$

$$\leqslant 2\left(\frac{1+P_r\exp\{\sqrt{\varepsilon/2}\}}{1-P_r}\right)^N \int_A^\infty \exp\{-\sqrt{u/2}\}du.$$

From this and (4) we obtain

$$|I_2| \leqslant C_1\int_A^\infty \exp\{-\sqrt{u/2}\}du,$$

where C_1 is some positive constant. It is clear that the last integral can be made arbitrarily small by choosing A.

Now we need to estimate the integral I_3. If $\varepsilon < |u| \leqslant \pi/d$, then there exists such $b > 0$ that $|\varphi(u)| \leqslant e^{-b}$. This, (2) and (4) imply that

$$|I_3| \leqslant 2\pi N^2(dB)^{-1}\left(\frac{e^{-b}+P_r}{1-P_r}\right)^N \to 0.$$

Lemma 1 is proved.

Now we consider the behaviour of $\nu_N^{(r)}$ as $n \to \infty$, $n/N^2 \to \infty$. By Lemma 1 this domain variation of N and n is the domain of large deviations. Since $\nu_N^{(r)}$ depends on the parameter r it follows that we need to prove the analogue of Theorem 1.3.7 under the condition of an array scheme.

Lemma 2. *Let $n \to \infty$ in such a way that $n/N^2 \to \infty$, r takes values which are divided by d and $r = n - zB^{-1}N^2$, where z is a fixed positive number. Then for n such that $\mathbf{P}\{\nu_N^{(r)} = N+n\} > 0$.*

$$\mathbf{P}\{\nu_N^{(r)} = N+n\} = \frac{dN}{2\pi n^{3/2}\sqrt{B}} \int_z^\infty (y^3 \exp\{1/y\})^{-1/2} dy (1+o(1)).$$

Proof. We set

$$\gamma = (N^2/n)^{1/6}. \tag{9}$$

It is easy to see that

$$\mathbf{P}\{\nu_N^{(r)} = N+n\} = P_{N1}(n) + NP_{N2}(n) + P_{N3}(n), \tag{10}$$

where

$$\begin{aligned}
P_{N1}(n) &= \mathbf{P}\{\nu_N^{(r)} = N+n;\ \nu_{[r]}^{(i)} \leqslant \gamma(N+n)d^{-1}+1,\ i=1,\dots,N\},\\
P_{N2}(n) &= \mathbf{P}\{\nu_N^{(r)} = N+n;\ \nu_{[r]}^{(i)} \leqslant \gamma(N+n)d^{-1}+1,\\
&\qquad i=1,\dots,N-1,\ \nu_{[r]}^{(N)} > \gamma(N+n)d^{-1}+1\},\\
P_{N3}(n) &= \mathbf{P}\{\nu_N^{(r)} = N+n;\ \bigcup_{i\neq j}(\nu_{[r]}^{(i)} > \gamma(N+n)d^{-1}+1,\\
&\qquad \nu_{[r]}^{(j)} > \gamma(N+n)d^{-1}+1)\}.
\end{aligned}$$

Estimating one after another these probabilities we will see that the second term gives the main contribution to the sum (10).

Firstly we estimate $P_{N1}(n)$. Let

$$R(w) = \sum_{k\leqslant\gamma(N+n)d^{-1}} \exp\{w(kd+1)\}\,\mathbf{P}\{\nu_{[r]}^{(1)} = kd+1\}. \tag{11}$$

Note that by (1.6)

$$\begin{aligned}
\sum_{k\leqslant l}(kd+1)\,\mathbf{P}\{\nu_{[r]}^{(1)} = kd+1\} &\leqslant (1-\mathbf{P}\{\nu^{(1)} > r+1\})^{-1}\\
&\quad\times\sum_{k\leqslant l}(kd+1)\,\mathbf{P}\{\nu^{(1)} = kd+1\}.
\end{aligned}\tag{12}$$

Since

$$\sum_{k\leqslant l}(kd+1)\,\mathbf{P}\{\nu^{(1)} = kd+1\}$$

$$= \sum_{k\leqslant\sqrt{l}}(kd+1)\,\mathbf{P}\{\nu^{(1)} = kd+1\} + \sum_{\sqrt{l}<k\leqslant l}(kd+1)\,\mathbf{P}\{\nu^{(1)} = kd+1\},$$

by using Lemma 1.3.12 to estimate the second term of the last expression, we obtain that for sufficiently large l

$$\begin{aligned}\sum_{k\leqslant l}(kd+1)\,\mathbf{P}\{\nu^{(1)}=kd+1\} &\leqslant d\sqrt{l}+1+C_1\sum_{k\leqslant l}(kd+1)^{-1/2}\\ &< C_2\sqrt{l}+C_3\int_0^l (yd+1)^{-1/2}dy \leqslant C_4\sqrt{l},\end{aligned} \tag{13}$$

where the symbols $C_1, C_2, \dots$ here and below denote positive constants. Using (12), (13) and the relation $\mathbf{P}\{\nu^{(1)} > r+1\} \to 0$, which is true as $r \to \infty$, we get that for sufficiently large l, r

$$\sum_{k\leqslant l}(kd+1)\,\mathbf{P}\{\nu_{[r]}^{(1)}=kd+1\} \leqslant C_4\sqrt{l}. \tag{14}$$

Reasoning by analogy it is not hard to see that

$$\sum_{k>l}\mathbf{P}\{\nu_{[r]}^{(1)}=kd+1\} < C_5\int_l^\infty y^{-3/2}dy = 2C_5 l^{-1/2}. \tag{15}$$

From the relation $N^2/n \to 0$ and (9) we obtain $(\gamma n)^{-1/2} = o(N^{-1})$. Using this and taking into account (11), (14), (15), (1.6) and the equation $e^y = 1+\delta(y)$, where $0 \leqslant y \leqslant 1$, $\delta(y) \leqslant 2y$, we obtain

$$R(1/(\gamma(N+n))) = 1 + o(N^{-1}). \tag{16}$$

Let $\nu_1(\gamma), \dots, \nu_N(\gamma)$ be auxiliary independent identically distributed random variables such that

$$\begin{aligned}&\mathbf{P}\{\nu_1(\gamma)=kd+1\}\\ &= \exp\{(kd+1)/(\gamma(N+n))\}\,\mathbf{P}\{\nu_{[r]}^{(1)}=kd+1\}/R(1/(\gamma(N+n))),\end{aligned}$$

where $k \leqslant \gamma(N+n)d^{-1}$. Let $\zeta_N(\gamma) = \nu_1(\gamma)+\dots+\nu_N(\gamma)$. Clearly

$$P_{N1}(n) = R^N(1/(\gamma(N+n)))e^{-1/\gamma}\,\mathbf{P}\{\zeta_N(\gamma)=N+n\}. \tag{17}$$

Now we prove that for sufficiently large n

$$\mathbf{P}\{\zeta_N(\gamma)=N+n\} \leqslant C_6 N^{-2}. \tag{18}$$

By $\varphi(u)$, $\varphi_r(u)$, $\varphi_{\gamma,r}(u)$ denote the characteristic functions of the random variables $\nu^{(1)}$, $\nu_{[r]}^{(1)}$, $\nu_1(\gamma)$, respectively. By the inversion formula

$$\begin{aligned}&\mathbf{P}\{\zeta_N(\gamma)=N+n\}\\ &= \frac{dB}{2\pi N^2}\int_{-\pi N^2/(dB)}^{\pi N^2/(dB)} \exp\left\{-\frac{iBu(N+n)}{N^2}\right\}\left(\varphi_{\gamma,r}\left(\frac{Bu}{N^2}\right)\right)^N du.\end{aligned} \tag{19}$$

We consider the expression

$$|\varphi_{\gamma,r}(u)|^N = \left|\frac{R(1/(\gamma(N+n))+iu)}{R(1/(\gamma(N+n)))}\right|^N. \tag{20}$$

If $k \leqslant \gamma(N+n)d^{-1}$ then $\exp\{(kd+1)/(\gamma(N+n))\} \leqslant 1 + 2(kd+1)/(\gamma(N+n))$; therefore

$$\begin{aligned} &|R(1/(\gamma(N+n)) + iu)| \\ &= \Big| \sum_{k \leqslant \gamma(N+n)d^{-1}} \exp\{(1/(\gamma(N+n)) + iu)(kd+1)\} \\ &\times \mathbf{P}\{\nu_{[r]}^{(1)} = kd+1\}\Big| \leqslant \Big| \sum_{k \leqslant \gamma(N+n)d^{-1}} \exp\{iu(kd+1)\} \\ &\times \mathbf{P}\{\nu_{[r]}^{(1)} = kd+1\}\Big| + 2(\gamma(N+n))^{-1} \sum_{k \leqslant \gamma(N+n)d^{-1}} (kd+1)\,\mathbf{P}\{\nu_{[r]}^{(1)} = kd+1\}. \end{aligned} \tag{21}$$

Using (14) we obtain

$$\begin{aligned} &(\gamma(N+n))^{-1} \sum_{k \leqslant \gamma(N+n)d^{-1}} (kd+1)\,\mathbf{P}\{\nu_{[r]}^{(1)} = kd+1\} \\ &\leqslant C_7(\gamma(N+n))^{-1/2} = o(N^{-1}). \end{aligned}$$

From this and (21) we get

$$\begin{aligned} &|R(1/(\gamma(N+n)) + iu)| \\ &\leqslant \left| \sum_{k \leqslant \gamma(N+n)d^{-1}} \exp\{iu(kd+1)\}\,\mathbf{P}\{\nu_{[r]}^{(1)} = kd+1\} \right| + o(N^{-1}) \\ &\leqslant |\varphi_r(u)| + \sum_{k > \gamma(N+n)d^{-1}} \mathbf{P}\{\nu_{[r]}^{(1)} = kd+1\} + o(N^{-1}). \end{aligned} \tag{22}$$

By (15)

$$\sum_{k > \gamma(N+n)d^{-1}} \mathbf{P}\{\nu_{[r]}^{(1)} = kd+1\} < C_8(\gamma(N+n))^{-1/2} = o(N^{-1}); \tag{23}$$

therefore, from (22) we obtain

$$|R(1/(\gamma(N+n)) + iu)| \leqslant |\varphi_r(u)| + o(N^{-1}). \tag{24}$$

The relation (3.1) gives us that for any u the inequality $|\varphi(u)| \geqslant C_9$ is true; therefore (4.1) and the condition $P_r \to 0$ imply that $|\varphi_r(u)| \geqslant C_{10}$. From this, (16), (20) and (24) we see that for any fixed u

$$|\varphi_{\gamma,r}(u)|^N \leqslant C_{11}|\varphi_r(u)|^N. \tag{25}$$

Now estimate $\varphi_r(u)$. From Lemma 1.3.3 we find that if $|u| \leqslant \varepsilon$ and ε is sufficiently small then $|\varphi(u)| \leqslant \exp\{-\sqrt{|u|/(2B)}\}$. From this we obtain that as $|u| \leqslant \varepsilon N^2$

$$|\varphi(u/N^2)| \leqslant \exp\{-\sqrt{|u|/(2BN^2)}\}. \tag{26}$$

As is well known, the inequality

$$|\varphi(u)| \leqslant e^{-C_{12}} \tag{27}$$

is valid if $0 < \varepsilon \leqslant |u| \leqslant \pi/d$. From (15) we see that for sufficiently large n

$$P_r = \sum_{k=1}^{\infty} \mathbf{P}\{\nu_{[r]}^{(1)} = r + kd + 1\} \leqslant C_{13} n^{-1/2}; \tag{28}$$

therefore, from (4.1), (27) and (28) we get that if $|u| \leqslant \varepsilon N^2$ then

$$\begin{aligned} \left|\varphi_r\left(\frac{Bu}{N^2}\right)\right|^N &\leqslant C_{14} \exp\{-\sqrt{|u|/2}\} \left(1 + \frac{C_{15} \exp\{\sqrt{\varepsilon B/2}\}}{\sqrt{n}}\right)^N \\ &\leqslant C_{16} \exp\{-\sqrt{|u|/2}\} \end{aligned} \tag{29}$$

and if $\varepsilon N^2 < |u| \leqslant \pi N^2/(dB)$ then

$$|\varphi_r(Bu/N^2)|^N \leqslant C_{17}(e^{-C_{12}} + C_{13} n^{-1/2})^N \leqslant C_{18} e^{-C_{12} N}. \tag{30}$$

We divide the integral (19) into the sum of two integrals with the domains of integration defined by the inequalities $|u| \leqslant \varepsilon N^2$ and $\varepsilon N^2 < |u| \leqslant \pi N^2/dB$. Then from (25), (29) and (30) we obtain

$$\mathbf{P}\{\zeta_N(\gamma) = N + n\} \leqslant C_{19} N^{-2} \int_0^{\infty} e^{-\sqrt{u/2}} du + C_{20} N^{-2}(N^2 e^{-C_{12} N}).$$

From this we find the estimation (18). Using (9), (16)–(18) we get

$$P_{N1}(n) \leqslant C_{21} N^{-2} e^{-1/\gamma} = o(N/n^{3/2}). \tag{31}$$

Now estimate $P_{N2}(n)$. It is clear that

$$\begin{aligned} P_{N2}(n) = &\sum_{n-r-1 \leqslant k < n-\gamma(N+n)d^{-1}-1} \mathbf{P}\{\nu_{[r]}^{(N)} = n - k\} \\ &\times \mathbf{P}\{\nu_{[r]}^{(1)} + \cdots + \nu_{[r]}^{(N-1)} = N + k,\ \nu_{[r]}^{(i)} \leqslant \gamma(N+n)d^{-1} + 1,\ i = 1, \ldots, N-1\}. \end{aligned} \tag{32}$$

Since

$$\begin{aligned} &\mathbf{P}\{\nu_{[r]}^{(1)} + \cdots + \nu_{[r]}^{(N-1)} = N + k,\ \nu_{[r]}^{(i)} \leqslant \gamma(N+n)d^{-1} + 1,\ i = 1, \ldots, N-1\} \\ &= \mathbf{P}\{\nu^{(1)} + \cdots + \nu^{(N-1)} = N + k,\ \nu^{(i)} \leqslant \gamma(N+n)d^{-1} + 1,\ i = 1, \ldots, N-1 | \\ &\qquad \nu^{(i)} \leqslant r + 1,\ i = 1, \ldots, N-1\} = \left[(1 - P_{\gamma(N+n)d^{-1}})/(1 - P_r)\right]^{N-1} \\ &\qquad \times \mathbf{P}\{\nu_{[\gamma(N+n)d^{-1}]}^{(1)} + \cdots + \nu_{[\gamma(N+n)d^{-1}]}^{(N-1)} = N + k\}, \end{aligned}$$

from (32) it follows that

$$\begin{aligned} P_{N2}(n) &= \left[(1 - P_{\gamma(N+n)d^{-1}})/(1 - P_r)\right]^{N-1} \\ &\times \sum_{n-r-1 \leqslant k < n-\gamma(N+n)d^{-1}-1} \mathbf{P}\{\nu_{[r]}^{(N)} = n - k\}\, \mathbf{P}\{\nu_{[\gamma(N+n)d^{-1}]}^{(1)} \\ &\qquad + \cdots + \nu_{[\gamma(N+n)d^{-1}]}^{(N-1)} = N + k\}. \end{aligned}$$

This together with (15), (23) implies that

$$P_{N2}(n) = (1+o(1)) \sum_{n-r-1\leqslant k<n-\gamma(N+n)d^{-1}-1} \mathbf{P}\{\nu_{[r]}^{(N)} = n-k\} \times \mathbf{P}\{\nu^{(1)}_{[\gamma(N+n)d^{-1}]} + \cdots + \nu^{(N-1)}_{[\gamma(N+n)d^{-1}]} = N+k\}. \tag{33}$$

As we have seen above the inequality, $|\varphi(u)| \geqslant C_9$ is valid for any u. Therefore from (4.1), (9), (15) and Lemma 1.3.3 we obtain by analogy with (3.39)

$$\left(\varphi_{\gamma(N+n)d^{-1}}(uB/N^2)\right)^{N-1} \to \exp\{\sqrt{|u|}(1-iu/|u|)\}. \tag{34}$$

Keeping in mind that the last expression is the characteristic function of the stable distribution with the parameter $\alpha = 1/2$ we derive from (34) that for $n \to \infty$ and any fixed y

$$\mathbf{P}\{\nu^{(1)}_{[\gamma(N+n)d^{-1}]} + \cdots + \nu^{(N-1)}_{[\gamma(N+n)d^{-1}]} \leqslant yN^2B^{-1}\} \to (2\pi)^{-1/2}\int_0^y (x^3\exp\{1/x\})^{-1/2}dx. \tag{35}$$

Let us prove that for sufficiently large n

$$\mathbf{P}\{\nu^{(1)}_{[\gamma(N+n)d^{-1}]} + \cdots + \nu^{(N-1)}_{[\gamma(N+n)d^{-1}]} > \gamma^{-1}N^2\} \leqslant C_{22}\gamma^{1/2}. \tag{36}$$

It is easy to see that

$$\begin{aligned}
&\mathbf{P}\{\nu^{(1)}_{[\gamma(N+n)d^{-1}]} + \cdots + \nu^{(N-1)}_{[\gamma(N+n)d^{-1}]} > \gamma^{-1}N^2\} \leqslant (N-1)\\
&\times \mathbf{P}\{\nu^{(1)}_{[\gamma(N+n)d^{-1}]} > \gamma^{-1}N^2+1\} + \mathbf{P}^{N-1}\{\nu^{(1)}_{[\gamma(N+n)d^{-1}]} \leqslant \gamma^{-1}N^2+1\}\\
&\times \mathbf{P}\{\nu^{(1)}_{[\gamma(N+n)d^{-1}]} + \cdots + \nu^{(N-1)}_{[\gamma(N+n)d^{-1}]} \geqslant \gamma^{-1}N^2 |\\
&|\nu^{(i)}_{[\gamma(N+n)d^{-1}]} \leqslant \gamma^{-1}N^2+1,\ i=1,\ldots,N-1\} \leqslant (N-1)\\
&\times \mathbf{P}\{\nu^{(1)}_{[\gamma(N+n)d^{-1}]} > \gamma^{-1}N^2+1\} + \left[(1-P_{\gamma^{-1}N^2})/(1-P_{\gamma(N+n)d^{-1}})\right]^{N-1}\\
&\times \mathbf{P}\{\nu^{(1)}_{[\gamma^{-1}N^2]} + \cdots + \nu^{(N-1)}_{[\gamma^{-1}N^2]} > \gamma^{-1}N^2\}.
\end{aligned} \tag{37}$$

We note that

$$\mathbf{P}\{\nu^{(1)}_{[\gamma(N+n)d^{-1}]} > \gamma^{-1}N^2+1\} = \sum_{k>\gamma^{-1}N^2d^{-1}} \mathbf{P}\{\nu^{(1)}_{[\gamma(N+n)d^{-1}]} = kd+1\} \leqslant C_{23}\,\mathbf{P}\{\nu^{(1)} > \gamma^{-1}N^2\}.$$

From this and (15) we obtain

$$(N-1)\,\mathbf{P}\{\nu^{(1)}_{[\gamma(N+n)d^{-1}]} > \gamma^{-1}N^2+1\} \leqslant C_{24}\gamma^{1/2}. \tag{38}$$

By (13) for sufficiently large n

$$\mathbf{E}\nu^{(N-1)}_{[\gamma^{-1}N^2]} \leqslant C_{25}\sum_{k<\gamma^{-1}N^2}(kd+1)\,\mathbf{P}\{\nu^{(1)} = kd+1\} \leqslant C_{25}\gamma^{-1/2}N.$$

Taking into account the Chebyshov inequality we obtain

$$\begin{aligned}\mathbf{P}\{\nu^{(1)}_{[\gamma^{-1}N^2]}+\cdots+\nu^{(N-1)}_{[\gamma^{-1}N^2]}>\gamma^{-1}N^2\}\\ \leqslant \gamma N^{-1}\,\mathbf{E}\,\nu^{(1)}_{[\gamma^{-1}N^2]}\leqslant C_{26}\gamma^{1/2}.\end{aligned}\tag{39}$$

By (15) and (23)

$$\big[(1-P_{\gamma^{-1}N^2})/(1-P_{\gamma(N+n)d^{-1}})\big]^{N-1}=1+o(1);$$

therefore (37)–(39) give us the estimation (36).

Using (33) we represent the probability $P_{N2}(n)$ as the sum

$$P_{N2}(n)=S_1+S_2+S_3,\tag{40}$$

where

$$\begin{gathered}S_i=(1+o(1))\sum_{K_i}\mathbf{P}\{\nu^{(N)}_{[r]}=n-k\}\\ \times\,\mathbf{P}\{\nu^{(1)}_{[\gamma(N+n)d^{-1}]}+\cdots+\nu^{(N-1)}_{[\gamma(N+n)d^{-1}]}=N+k\},\\ K_1=\{k:\ n-r-1\leqslant k\leqslant\gamma^{-1}N^2\},\quad K_2=\{k:\ \gamma^{-1}N^2<k\leqslant n(1-\gamma^{1/6})\},\\ K_3=\{k:\ n(1-\gamma^{1/6})<k<n(1-\gamma)-\gamma Nd^{-1}-1\}.\end{gathered}$$

Using (28), (1.6) and Lemma 1.3.12 it is not hard to get that as $n-r<k\leqslant\gamma^{-1}N^2$ and $r=n-zB^{-1}N^2$

$$\begin{aligned}\mathbf{P}\{\nu^{(N)}_{[r]}=n-k\}=(1-P_r)^{-1}\,\mathbf{P}\{\nu^{(N)}=n-k\}\\ =d(2\pi Bn^3)^{-1/2}(1+o(1))\end{aligned}\tag{41}$$

uniformly in k. Therefore

$$\begin{aligned}S_1=(d+o(1))(2\pi Bn^3)^{-1/2}\sum_{K_1}\mathbf{P}\{\nu^{(1)}_{[\gamma(N+n)d^{-1}]}+\cdots+\nu^{(N-1)}_{[\gamma(N+n)d^{-1}]}\\ =N+k\}=(d+o(1))(2\pi Bn^3)^{-1/2}\,\mathbf{P}\{N+n-r-1\leqslant\nu^{(1)}_{[\gamma(N+n)d^{-1}]}\\ +\cdots+\nu^{(N-1)}_{[\gamma(N+n)d^{-1}]}\leqslant N+\gamma^{-1}N^2\}.\end{aligned}$$

Putting $r=n-zB^{-1}N^2$ from this and (35), (36) we obtain

$$S_1=(d+o(1))(2\pi n^{3/2}\sqrt{B})^{-1}\int_z^\infty y^{-3/2}e^{-1/(2y)}dy.\tag{42}$$

Let us prove that the relations $S_2=o(n^{-3/2})$, $S_3=o(n^{-3/2})$ are true. If $\gamma^{-1}N^2<k\leqslant n(1-\gamma^{1/6})$ then it is easy to see by analogy with the proof of (41)

$$\mathbf{P}\{\nu^{(N)}_{[r]}=n-k\}\leqslant C_{27}\,\mathbf{P}\{\nu^{(N)}=n-k\}\leqslant C_{28}(n\gamma^{1/6})^{-3/2}.$$

This implies that

$$S_2 \leqslant C_{29}(n\gamma^{1/6})^{-3/2} \sum_{\gamma^{-1}N^2<k\leqslant n(1-\gamma^{1/6})} \mathbf{P}\{\nu^{(1)}_{[\gamma(N+n)d^{-1}]}$$
$$+\cdots+\nu^{(N-1)}_{[\gamma(N+n)d^{-1}]} = N+k\} \leqslant C_{30}(n\gamma^{1/6})^{-3/2}$$
$$\times \mathbf{P}\{\nu^{(1)}_{[\gamma(N+n)d^{-1}]}+\cdots+\nu^{(N-1)}_{[\gamma(N+n)d^{-1}]} > \gamma^{-1}N^2\}.$$

Combining this and (36) we get $S_2 \leqslant C_{31}n^{-3/2}\gamma^{1/4}$; hence

$$S_2 = o(n^{-3/2}). \tag{43}$$

Now we consider S_3. If $n(1-\gamma^{1/6}) < k < n(1-\gamma) - \gamma N d^{-1} - 1$, then as before

$$\mathbf{P}\{\nu^{(N)}_{[r]} = n-k\} \leqslant C_{32}(n\gamma)^{-3/2}.$$

From this we obtain

$$S_3 \leqslant C_{32}(n\gamma)^{-3/2}\,\mathbf{P}\{\nu^{(1)}_{[\gamma(N+n)d^{-1}]}+\cdots+\nu^{(N-1)}_{[\gamma(N+n)d^{-1}]} > n(1-\gamma^{1/6})\}. \tag{44}$$

According to (14) $\mathbf{E}\,\nu^{(1)}_{[r]} \leqslant C_{33}\sqrt{r}$; therefore by the Chebyshov inequality

$$\mathbf{P}\{\nu^{(1)}_{[\gamma(N+n)d^{-1}]}+\cdots+\nu^{(N-1)}_{[\gamma(N+n)d^{-1}]} > n(1-\gamma^{1/6})\} \leqslant C_{34}N(\gamma/n)^{1/2}.$$

From this, (9) and (44) we see that $S_3 \leqslant C_{35}N/(n^2\gamma) = o(n^{-3/2})$. This together with (40), (42), (43) gives us that

$$P_{N2}(n) = (d+o(1))(2\pi n^{3/2}\sqrt{B})^{-1}\int_z^\infty (y^3\exp\{1/y\})^{-1/2}dy. \tag{45}$$

To estimate $P_{N3}(n)$ we note that

$$P_{N3}(n) = N(N-1)2^{-1}\sum_{k<(N+n)(1-2\gamma d^{-1})} \mathbf{P}\{\nu^{(1)}_{[r]}+\cdots+\nu^{(N-2)}_{[r]} = k\}$$
$$\times \mathbf{P}\{\nu^{(N-1)}_{[r]}+\nu^{(N)}_{[r]} = N+n-k,\ \nu^{(N-1)}_{[r]} > \gamma(N+n)d^{-1}+1,$$
$$\nu^{(N)}_{[r]} > \gamma(N+n)d^{-1}+1\}.$$

Therefore

$$P_{N3}(n) \leqslant C_{35}N^2\sum_{k<(N+n)(1-2\gamma d^{-1})} \mathbf{P}\{\nu^{(1)}_{[r]}+\cdots+\nu^{(N-2)}_{[r]} = k\}$$
$$\times\left(\sum_S \mathbf{P}\{\nu^{(N-1)}_{[r]} = s\}\,\mathbf{P}\{\nu^{(N)}_{[r]} = N+n-k-s\}\right), \tag{46}$$

where $S = \{s:\ \gamma(N+n)d^{-1}+1 < s < (N+n)(1-\gamma d^{-1})-k-1\}$. By Lemma 1.3.12

$$\mathbf{P}\{\nu^{(N-1)}_{[r]} = s\} \leqslant C_{36}(\gamma n)^{-3/2}$$

as $s > \gamma(N+n)d^{-1}$. Using (15) we obtain

$$\sum_S \mathbf{P}\{\nu_{[r]}^{(N-1)} = s\}\,\mathbf{P}\{\nu_{[r]}^{(N)} = N+n-k-s\}$$
$$\leqslant C_{36}(\gamma n)^{-3/2}\,\mathbf{P}\{\nu_{[r]}^{(N)} > \gamma(N+n)d^{-1}+1\} < C_{37}(\gamma n)^{-2}.$$

This and (46) implies that

$$P_{N3}(n) \leqslant C_{38}N^2(\gamma n)^{-2} \sum_{k<(N+n)(1-2\gamma d^{-1})} \mathbf{P}\{\nu_{[r]}^{(1)} + \cdots + \nu_{[r]}^{(N-2)} = k\}$$
$$\leqslant C_{38}N^2(\gamma n)^{-2} = o(N/n^{3/2}).$$

Lemma 2 is obtained from this and (10), (31), (45).

6. Proofs of the main results

In Section 1 the main theorems of Chapter 2 were formulated. Now we will prove these results.

Let the hypotheses of Theorem 1.1 be satisfied. By Lemma 2.2

$$NP_{r-d} \to \infty, \quad NP_{r+k} \to \gamma, \quad k = 0, d, 2d, \ldots, s-d. \tag{1}$$

Then

$$(1-P_r)^N \to 0, \quad (1-P_{r+k})^N \to e^{-\gamma}, \quad k = 0, d, 2d, \ldots, s-d. \tag{2}$$

By analogy with (2.3) it is not hard to obtain

$$\begin{aligned} P_{r+s} &= \sum_{i=1}^{\infty} \mathbf{P}\{\nu^{(1)} = r+s+i+1\} \\ &= \mathbf{P}\{\nu^{(1)} = r+s+l+1\}\left(1 + \sum_{i=l+1}^{\infty} \frac{\mathbf{P}\{\nu^{(1)} = r+s+i+1\}}{\mathbf{P}\{\nu^{(1)} = r+s+l+1\}}\right), \end{aligned} \tag{3}$$

where l is the least natural number such that $\mathbf{P}\{\nu^{(1)} = r+s+l+1\} > 0$. From (2.4) it follows that

$$\begin{aligned} &\mathbf{P}\{\nu^{(1)} = r+s+i+1\}/\,\mathbf{P}\{\nu^{(1)} = r+s+l+1\} \\ &= \lambda^{i-l}\,\mathbf{P}\{\nu_*^{(1)} = r+s+i+1\}/\left(F^{i-l}(\lambda)\,\mathbf{P}\{\nu_*^{(1)} = r+s+l+1\}\right), \end{aligned} \tag{4}$$

hence by virtue of (3)

$$P_{r+s} = \mathbf{P}\{\nu^{(1)} = r+s+l+1\}(1+o(1)). \tag{5}$$

From this and the condition $N\,\mathbf{P}\{\nu^{(1)} = r+s+1\} \to \gamma$ it is clear that $N\,\mathbf{P}\{\nu^{(1)} = r+s+l+1\} \to 0$. The last relation together with (5) yields

$$(1-P_{r+s})^N \to 1. \tag{6}$$

It is not hard to see that under the conditions of Theorem 1.1 the hypotheses of Lemma 3.2 are satisfied. Indeed, if r is fixed, then from (2.4) and the relation $N\,\mathbf{P}\{\nu^{(1)} = r+1\} \to \infty$ it follows that $N\lambda^r \to \infty$. Therefore the inequality $r \geqslant j+l$ implies $N\lambda^{j+l} \to \infty$. Using the first assertion of Lemma 2.1 we see that $\lambda/F(\lambda) =$

$o(1)$; hence from the relation $N\,\mathbf{P}\{\nu^{(1)} = r+1\} \to \infty$ we obtain for sufficiently large r

$$N\lambda^{j+l} \geqslant C_3 N(\lambda/F(\lambda))^{j+l} > N(\lambda/F(\lambda))^r,$$

where C_3 is a positive constant. From this and (2.4) we obtain that for sufficiently large N, n, r

$$N\lambda^{j+l} \geqslant N\,\mathbf{P}\{\nu^{(1)} = r+1\};$$

therefore the hypotheses of Lemma 3.2 are satisfied as $r \to \infty$. Using this lemma we obtain

$$\mathbf{P}\{\nu_N = N+n\} = (d+o(1))(2\pi N\sigma^2)^{-1/2}. \tag{7}$$

According to (1) the relation $NP_r \to \gamma$ is valid. Therefore by Lemma 4.2

$$\mathbf{P}\{\nu_N^{(r)} = N+n\} = (d+o(1))(2\pi N\sigma^2)^{-1/2}.$$

This together with (2), (6), (7) and Lemma 1.1 implies that

$$\mathbf{P}\{\eta \leqslant r-d-1\} \to 0, \quad \mathbf{P}\{\eta \leqslant r+k+1\} \to e^{-\gamma},$$
$$k = 0, d, 2d, \dots, s-d, \quad \mathbf{P}\{\eta \leqslant r+s+1\} \to 1.$$

Theorem 1.1 follows immediately from these relations.

By virtue of Lemma 2.3, under the conditions of Theorem 1.2 for a fixed integer k

$$(1-P_{r+kd})^N = \exp\{-\gamma\alpha^{k+1}(1-\alpha)^{-1}\}. \tag{8}$$

From the second assertion of Lemma 2.1 it follows that $\lambda \geqslant C > 0$. Therefore the hypotheses of Lemma 3.2 are satisfied. Clearly, Lemma 2.3 implies the conditions of Lemma 4.2. Using Lemmas 3.2 and 4.2 we obtain

$$\mathbf{P}\{\nu_N^{(r)} = N+n\}/\,\mathbf{P}\{\nu_N = N+n\} \to 1. \tag{9}$$

From this, Lemma 1.1 and (8) we deduce the assertion of Theorem 1.2.

To prove Theorem 1.3 we point out that by virtue of Lemma 2.5 for r which are divided by d and such that $r = (u+z)/\beta + O(1)$, where u and β are given by (1.13) and (1.12) respectively, z is a fixed number, the equality

$$(1-P_r)^N = e^{-e^{-z}}(1+o(1)) \tag{10}$$

holds. From Lemmas 3.2 and 4.2 it follows that under the conditions of Theorem 1.3 the relation (9) is valid. Therefore, Lemma 1.1 and (10) give us the assertion of Theorem 1.3.

Let us prove Theorem 1.4. If $Bn/N^2 \to \gamma$, r is divided by d and $r = zn + O(1)$, then from Lemma 2.6 we obtain that $NP_r \to E(0,z)$, where the function $E(u,z)$ is determined by (5.1). From this it follows that

$$(1-P_r)^N = e^{-E(0,z)}(1+o(1)). \tag{11}$$

By Lemma 3.3

$$\mathbf{P}\{\nu_N = N+n\} = dN(2\pi Bn^3\exp\{1/\gamma\})^{-1/2}(1+o(1)) \tag{12}$$

and by Lemma 5.1

$$\mathbf{P}\{\nu_N^{(r)} = N+n\} = \frac{dB}{N^2\sqrt{2\pi}} e^{E(0,z)} \sum_{k=0}^{\infty} \frac{(-1)^k}{k!} I_k(\gamma z, \gamma).$$

Then using (11), (12) Theorem 1.4 can be derived from Lemma 1.1.

To prove Theorem 1.5 it is sufficient to note that by Lemma 2.7 for r which are divided by d and such that $r = n - zB^{-1}N^2 + O(1)$, where z is a positive constant, the equality

$$(1 - P_r)^N = 1 + o(1) \tag{13}$$

is valid. By virtue of Lemma 3.4

$$\mathbf{P}\{\nu_N = N + n\} = dN(2\pi n^3 B)^{-1/2}(1 + o(1)), \tag{14}$$

and by Lemma 5.2

$$\mathbf{P}\{\nu_N^{(r)} = N + n\} = \frac{dN}{2\pi n^{3/2}\sqrt{B}} \int_z^\infty (y^3 \exp\{1/y\})^{-1/2} dy (1 + o(1)).$$

Now the validity of Theorem 1.5 follows from (13), (14) and Lemma 1.1.

In addition we find out some sufficient conditions for the series $r = r(N, n)$ for Theorem 1.1. Let $g(N, n, r) = Nq^r/r^{3/2}$, where $q = \lambda/F(\lambda)$, and let $k = [-(\ln N)/(d \ln q)]$; square brackets denote the integer part.

Theorem 1. *Let $N, n \to \infty$ in such a way that n takes values which are divided by d, $n/N \to 0$. Let also $r = kd$, s_1, s_2 be natural numbers divided by d and*

$$\begin{aligned} &\sum_K p_{k_1} p_{k_2} \cdots p_{k_{r+1}} > 0, \quad \sum_{K_{s_1}} p_{k_1} p_{k_2} \cdots p_{k_{r-s_1+1}} > 0, \\ &\sum_{K_{s_2}} p_{k_1} p_{k_2} \cdots p_{k_{r+s_2+1}} > 0, \quad \sum_{K_i} p_{k_1} p_{k_2} \cdots p_{k_{r+i+1}} = 0, \end{aligned} \tag{15}$$

where summation domains K, K_{s_1}, K_{s_2}, K_i contain the non-negative integers $k_1, \ldots, k_{r+s_2+1}$ such that

$$\begin{aligned} &K = \{k_1, \ldots, k_{r+1} : k_1 + \cdots + k_{r+1} = r\}, \\ &K_{s_1} = \{k_1, \ldots, k_{r-s_1+1} : k_1 + \cdots + k_{r-s_1+1} = r - s_1\}, \\ &K_{s_2} = \{k_1, \ldots, k_{r+s_2+1} : k_1 + \cdots + k_{r+s_2+1} = r + s_2\}, \\ &K_i = \{k_1, \ldots, k_{r+s_2} : k_1 + \cdots + k_{r+i+1} = r + i\}, \\ &i = -s_1 + 1, -s_1 + 2, \ldots, -1, 1, \ldots, s_2 - 1, \end{aligned}$$

and probabilities p_k, $k = 0, 1, \ldots$ are determined by (1.1). Let one of the conditions

1) *$g(N, n, r) \to \infty$,*
2) *$g(N, n, r)$ is limited,*
3) *there exists a series $r = r(N, n)$ such that $g(N, n, r)$ is limited*

hold. The following assertions are true:

1. *Under the condition 1) $N\,\mathbf{P}\{\nu^{(1)} = r + 1\} \to \infty$ and $N\,\mathbf{P}\{\nu^{(1)} = r + s_2 + 1\}$ is limited for fixed r and $N\,\mathbf{P}\{\nu^{(1)} = r + d + 1\}$ is limited for $r \to \infty$.*
2. *Under the conditions 2) or 3) $N\,\mathbf{P}\{\nu^{(1)} = r + 1\}$ is limited and $N\,\mathbf{P}\{\nu^{(1)} = r - s_1 + 1\} \to \infty$ for fixed r and $N\,\mathbf{P}\{\nu^{(1)} = r - d + 1\} \to \infty$ for $r \to \infty$.*

Proof. Using (15) we obtain

$$\begin{aligned}&\mathbf{P}\{\nu_*^{(1)} = r+1\} > 0, \quad \mathbf{P}\{\nu_*^{(1)} = r - s_1 + 1\} > 0,\\ &\mathbf{P}\{\nu_*^{(1)} = r+s_2+1\} > 0, \quad \mathbf{P}\{\nu_*^{(1)} = r+i+1\} = 0,\end{aligned} \tag{16}$$

where $i = -s_1+1, -s_1+2, \ldots, -1, 1, \ldots, s_2-1$. Let the condition 1) hold. If r is fixed then from (2.4) and (16) it follows that

$$N\,\mathbf{P}\{\nu^{(1)} = r+1\} = O(g(N,n,r)); \tag{17}$$

therefore $N\,\mathbf{P}\{\nu^{(1)} = r+1\} \to \infty$. By analogy $N\,\mathbf{P}\{\nu^{(1)} = r+s_2+1\} = O(g(N,n,r+s_2))$ and substituting kd for r in $g(N,n,r+s_2)$ we find that the expression $N\,\mathbf{P}\{\nu^{(1)} = r+s_2+1\}$ is limited.

Now consider the case $r \to \infty$. From (2.6) it follows that if r is sufficiently large, then $\mathbf{P}\{\nu^{(1)} = r+d+1\} > 0$; therefore $s_2 = d$. Using (17) we obtain $N\,\mathbf{P}\{\nu^{(1)} = r+1\} \to \infty$. From (2.4) and (2.6) we get $N\,\mathbf{P}\{\nu^{(1)} = r+d+1\} = O(Nq^{r+d}r^{-3/2})$; therefore $N\,\mathbf{P}\{\nu^{(1)} = r+d+1\} = O(q^d g(N,n,r))$. Taking the logarithm of the expression $q^d g(N,n,r)$ and taking into account the relations $r = kd$ and $q \to 0$ we find that the expression $N\,\mathbf{P}\{\nu^{(1)} = r+d+1\}$ is limited.

Under one of the conditions 2) or 3) from (15) we get that $r \geqslant s_1$ for fixed r and (17) implies that $N\,\mathbf{P}\{\nu^{(1)} = r+1\}$ is limited. Since $N\,\mathbf{P}\{\nu^{(1)} = r-s_1+1\} = O(q^{-s_1}g(N,n,r))$, it follows that $N\,\mathbf{P}\{\nu^{(1)} = r-s_1+1\} \to \infty$. If $r \to \infty$, then from (17) we obtain that $N\,\mathbf{P}\{\nu^{(1)} = r+1\}$ is limited but $N\,\mathbf{P}\{\nu^{(1)} = r-d+1\} \to \infty$. This together with the relation $N\,\mathbf{P}\{\nu^{(1)} = r-d+1\} = O(q^{-d}g(N,n,r))$ implies the assertion of Theorem 1.

7. Additions and references

Chapter 2 is based on the article [65]. Limit behaviour of the maximum tree size was first studied in [53,54] where this problem was solved for the $\mathfrak{F}'_{N,n}$ forest class (example 1.1.1) under uniform probability distribution.

Lemma 2.4 was given in [6]. Results of Sections 3 and 4 were proved in [65,67]. The proofs of Lemmas 3.3, 3.4, 5.1 and 5.2 make use of the known (e. g. see [77]) fact that the expression $\exp\{-\sqrt{-2iu}\}$ is the characteristic function of the stable law with the exponent $\alpha = 1/2$ and the density $(2\pi x^3 \exp\{1/x\})^{-1/2}$.

Theorem 6.1 is analogous to Theorem 4.1 from [10]. Results for the maximum tree size in a random forest of non-rooted trees with labelled vertices under uniform probability distribution on a set of such forests were obtained in [7]. In papers [10,11,47], conditions for emergence of a gigantic component in a random forest were revealed, and possible uses of the results about non-rooted trees in the study of random graph evolution demonstrated. Note that [10] also gives limit distributions for the size of the k-th largest tree, $k = 1, 2, \ldots$

A similar problem for the forests considered in the present book was studied in [13,14]. Let $\nu_{(1)}(\mathfrak{F}), \ldots, \nu_{(N)}(\mathfrak{F})$ be the set of order tree sizes $\nu_1(\mathfrak{F}), \ldots, \nu_N(\mathfrak{F})$ such that $\nu_{(1)}(\mathfrak{F}) \leqslant \nu_{(2)}(\mathfrak{F}) \leqslant \ldots \leqslant \nu_{(N)}(\mathfrak{F})$. Also, let $h = 0, 1, 2, \ldots$

Theorem 1. *Under the conditions of Theorem* 1.1

$$\mathbf{P}\{\nu_{(N-h)}(\mathfrak{F}) = r + 1\} = e^{-\gamma} \sum_{g=0}^{h} \gamma^g / g!,$$

$$\mathbf{P}\{\nu_{(N-h)}(\mathfrak{F}) = r + s + 1\} = 1 - e^{-\gamma} \sum_{g=0}^{h} \gamma^g / g!$$

Theorem 2. *Under the conditions of Theorem* 1.2 *for any fixed integer k*

$$\mathbf{P}\{\nu_{(N-h)}(\mathfrak{F}) \leqslant r + kd + 1\} \to \exp\left\{ -\frac{\gamma \alpha^{k+1}}{1-\alpha} \sum_{g=0}^{h} \frac{1}{g!} \left(\frac{\gamma \alpha^{k+1}}{1-\alpha} \right)^g \right\}.$$

Theorem 3. *Under the conditions of Theorem* 1.3

$$\mathbf{P}\{\beta \nu_{(N-h)}(\mathfrak{F}) - u \leqslant z\} \to e^{-e^{-z}} \sum_{g=0}^{h} e^{-gz} / g!$$

Theorem 4. *Under the conditions of Theorem* 1.4

$$\mathbf{P}\{\nu_{(N-h)}(\mathfrak{F})/n \leqslant z\} \to$$

$$\to \exp\{1/(2\gamma)\} \gamma^{3/2} \sum_{k=0}^{\infty} \frac{(-1)^k}{k!} \sum_{g=0}^{h} \frac{1}{g!} I_{k+g}(\gamma z, \gamma).$$

Theorem 5. *Under the conditions of Theorem* 1.5

$$\mathbf{P}\{\nu_{(N-h)}(\mathfrak{F}) \leqslant zN^2\} \to e^{-E(z)} \sum_{g=0}^{h-1} E^g(z)/g!,$$

where $E(z) = \sqrt{2/(z\pi B)}$, $h = 1, 2, \ldots$

Theorems 1.1–1.5 and 1–5 show that in a random forest the gigantic tree with almost all n non-root vertices arises only if $N, n \to \infty$, $n/N^2 \to \infty$.

CHAPTER 3

LIMIT DISTRIBUTIONS OF THE NUMBER OF TREES OF A GIVEN SIZE

1. Problem statement and summary of results

Let us remind the reader of the main notations introduced in Chapter 1 for the set of forests $\mathfrak{F}_{N,n}$. Let $\nu_1(\mathfrak{F}), \dots, \nu_N(\mathfrak{F})$ be the sizes of trees with roots $1, \dots, N$ in a forest from $\mathfrak{F}_{N,n}$. The set $\mathfrak{F}_{N,n}$ is connected with some Galton–Watson process G consisting of N independent processes $G^{(1)}, \dots, G^{(N)}$ beginning with one particle. Let the number of offspring of a particle in the process G have the distribution

$$p_k(\lambda) = \lambda^k p_k / F(\lambda), \quad k = 0, 1, 2, \dots, \tag{1}$$

where $0 < \lambda \leqslant 1$ and the probabilities p_k determine the discrete distribution (2.1.1) of the random variable ξ with the maximum span d and the generating function

$$F(z) = \sum_{k=0}^{\infty} p_k z^k. \tag{2}$$

Also, the set of values ξ with non-zero probability includes null and differs from $\{0, 1\}$. We will also remind the reader that $\mathbf{E}\xi = 1$, $\mathbf{D}\xi = B$.

Let $\nu_{(r)}^{(1)}, \dots \nu_{(r)}^{(N)}$ be independent identically distributed random variables such that

$$\mathbf{P}\{\nu_{(r)}^{(i)} = k\} = \mathbf{P}\{\nu^{(i)} = k |\, \nu^{(i)} \neq r + 1\}, \tag{3}$$

where $i = 1, \dots, N$, $r = 0, 1, \dots$, and $\nu^{(i)}$ is the total progeny of the process $G^{(i)}$, $\nu^{(1)} + \dots + \nu^{(N)} = \nu_N$.

Let $\nu_{(r)}^{(1)}$ have the maximum span d. It is clear that, in another case, the results of this Chapter can be easily transformed.

We denote by $\mu_r(\mathfrak{F})$ the number of trees from $\mathfrak{F}_{N,n}$ with r non-root vertices, $r = 0, 1, \dots, n$. The relation (2.1.5) is an example of the equation (1.2.1). Therefore the conditions of the generalized allocation scheme are valid and from Lemma 1.2.1 we obtain the next assertion.

Lemma 1. *For any* λ, $0 < \lambda \leqslant 1$ *and* n *such that* $\mathbf{P}\{\nu_N = N + n\} > 0$

$$\mathbf{P}\{\mu_r(\mathfrak{F}) = k\} = \binom{N}{k} q_r^k(\lambda)(1 - q_r(\lambda))^{N-k} \frac{\mathbf{P}\{\zeta_{N-k}^{(r)} = N + n - k(r+1)\}}{\mathbf{P}\{\nu_N = N + n\}},$$

where $q_r(\lambda) = \mathbf{P}\{\nu^{(1)} = r + 1\}$, $\zeta_{N-k}^{(r)} = \nu_{(r)}^{(1)} + \dots + \nu_{(r)}^{(N-k)}$.

The asymptotics of the probabilities

$$\binom{N}{k} q_r^k(\lambda)(1-q_r(\lambda))^{N-k}$$

is well known and the limit behaviour of the probability $\mathbf{P}\{\nu_N = N+n\}$ is studied in Section 2.3. Therefore Lemma 1 shows that to investigate the behaviour of $\mu_r(\mathfrak{F})$, it suffices to consider the probability $\mathbf{P}\{\zeta_{N-k}^{(r)} = N+n-k(r+1)\}$ in various domains of variation of N, n, r. This problem is considered in Sections 2, 3 and in Section 4 we will prove the limit theorems for $\mu_r(\mathfrak{F})$.

Let the parameter $\lambda = \lambda(N, n)$ be determined by the relation

$$\frac{\lambda F'(\lambda)}{F(\lambda)} = \frac{n}{N+n}. \tag{4}$$

We denote by j the least positive integer such that $p_j > 0$ and let l be the least natural number not divided by j for which $p_{j+l} > 0$; if such l does not exist, we put $l = 0$.

Theorem 1. *Let $N, n \to \infty$ in such a way that $n/N \to 0$, $N\lambda^{j+l} \to \infty$ and let n be divided by d. Then for $2 \leqslant r \neq j$*

$$\mathbf{P}\{\mu_r(\mathfrak{F}) = k\} = (k!)^{-1}(Nq_r(\lambda))^k \exp\{-Nq_r(\lambda)\}(1+o(1))$$

uniformly in the integers k such that $(k - Nq_r(\lambda))/\sqrt{Nq_r(\lambda)}$ lies in any finite interval.

Remark 1. Using (2.2.4) it is easy to get

$$q_r(\lambda) = \lambda^r F^{-(r+1)}(\lambda) \sum p_0^{k_0} p_1^{k_1} \cdots p_r^{k_r}, \tag{5}$$

where $k_0, k_1, \dots, k_r$ are non-negative integers and the summation domain is

$$\{k_0, k_1, \dots, k_r : \ k_0 + \cdots + k_r = r+1\}.$$

Let

$$\sigma_{rr}^2 = q_r(\lambda)\left(1 - q_r(\lambda) - \frac{(a-r-1)^2}{\sigma^2} q_r(\lambda)\right), \tag{6}$$

where (see (2.3.2))

$$a = \mathbf{E}\,\nu^{(1)} = 1/(1-m), \quad \sigma^2 = \mathbf{D}\,\nu^{(1)} = B_\lambda/(1-m)^3, \tag{7}$$

m and B_λ are mathematical expectation and variance of the distribution (1) respectively and by (2.2.1)

$$m = m(\lambda) = (\lambda F'(\lambda))/F(\lambda). \tag{8}$$

Theorem 2. *Let $N, n \to \infty$ in such a way that $n/N \geqslant C > 0$, n be divided by d and $F'''(1) < \infty$ as $n/N \to \infty$, $n/N^2 \to 0$. Then*

$$\mathbf{P}\{\mu_r(\mathfrak{F}) = k\} = (\sigma_{rr}\sqrt{2\pi N})^{-1} e^{-u_r^2/2}(1+o(1))$$

uniformly in the integers k such that $u_r = (k - Nq_r(\lambda))/(\sigma_{rr}\sqrt{N})$ lies in any finite interval. For $n/N^2 \geqslant C > 0$, the values $q_r(\lambda)$ and σ_{rr}^2 can be replaced by $q_r(1)$ and $q_r(1)(1-q_r(1))$, respectively.

Let s denote the least natural number such that $p_{j+s} > 0$; if there exists no such s, we put $s = 0$. Thus, s differs from l in that s can be divided by j.

Theorem 3. *Let $N, n \to \infty$ in such a way that $n/N \to 0$, and let n be divided by d and* $\min(N\lambda^r, N\lambda^{\omega(r)}) \to \infty$, *where*

$$\begin{aligned}
&\omega(0) = 2j + l; \\
&\omega(1) = 3 && \text{for } j = 1; \\
&\omega(1) = j + l && \text{for } j > 1; \\
&\omega(r) = j + l && \text{for } r \geqslant 2, r \neq j; \\
&\omega(r) = \max\{\min(2j, j+s), j+l\} && \text{for } r = j \geqslant 2.
\end{aligned}$$

Then the assertion of Theorem 2 is valid.

Theorem 4. *Let $N, n, r \to \infty$ in such a way that $N\lambda^j \to \infty$, n be divided by d and $F'''(1) < \infty$ as $n/N \to \infty$, $n/N^2 \to 0$. Then*

$$\mathsf{P}\{\mu_r(\mathfrak{F}) = k\} = \frac{1}{k!}(Nq_r(\lambda))^k e^{-Nq_r(\lambda)}(1 + o(1))$$

uniformly in integers k such that $(k - Nq_r(\lambda))/\sqrt{Nq_r(\lambda)}$ lies in any finite interval. For $n/N^2 \geqslant C > 0$, the values $q_r(\lambda)$ can be replaced by $q_r(1)$.

2. The convergence of the sum of auxiliary random variables to the normal law

In this Section we will get the limit distributions of the sum $\zeta_S^{(r)} = \nu_{(r)}^{(1)} + \cdots + \nu_{(r)}^{(S)}$ as $N, n, S \to \infty$, $n/N^2 \to 0$. Let

$$\varphi^{(r)}(u) = \mathsf{E}\exp\{iu\nu_{(r)}^{(1)}\} = \frac{\varphi(u) - q_r(\lambda)\exp\{iu(r+1)\}}{1 - q_r(\lambda)}, \tag{1}$$

where $\varphi(u)$ is the characteristic function of the random variable $\nu^{(1)}$. Let the parameter $\lambda = \lambda(N, n)$ be determined by the relation (1.4). We denote also

$$a_r = \mathsf{E}\,\nu_{(r)}^{(1)}, \quad \sigma_r^2 = \mathsf{D}\,\nu_{(r)}^{(1)}.$$

It is not hard to check that

$$\begin{aligned}
a_r &= (a - (r+1)q_r(\lambda))/(1 - q_r(\lambda)), \\
\sigma_r^2 &= \frac{\sigma^2}{(1 - q_r(\lambda))^2}(1 - q_r(\lambda) - \frac{(a - r - 1)^2}{\sigma^2} q_r(\lambda)).
\end{aligned} \tag{2}$$

Let $\varphi_S^{(r)}(u)$ be the characteristic function of the random variable $(\zeta_S^{(r)} - Sa_r)/(\sigma_r\sqrt{S})$. The following assertion is valid.

Lemma 1. *Let $r \geqslant 2$, $r \neq j$, $N, n \to \infty$ in such a way that $n/N^2 \to 0$, $N\lambda^j \to \infty$, $F'''(1) < \infty$ if $n/N \to \infty$. Then for $S = N(1 - q_r(\lambda))(1 + o(1))$*

$$\varphi_S^{(r)}(u) \to e^{-u^2/2}$$

uniformly in u lying in any finite interval. This assertion remains true for $r = 1$, $j > 1$; for $r \to \infty$ and also in such cases:

1) *$r = j = 2$ or $r = 1, j = 1$ or $r = 0$ (except $j = 2$, $p_3 = 0$, $p_4 > 0$), if $N\lambda^\omega \to \infty$, where $\omega = \min(2j, j + s)$;*

2) *$r = 0, j = 2, p_3 = 0, p_4 > 0$, if $N\lambda^2 \to \infty$.*

Proof. To prove Lemma 1 it is important to know the evident expression of $(\ln\varphi^{(r)}(u))'''_u$. Using (2.3.1) and (1) we can obtain that

$$
\begin{aligned}
&(\ln\varphi^{(r)}(u))'''_u = -i(\varphi(u) - q_r(\lambda)\exp\{iu(r+1)\})^{-3}(1\\
&-e^{iu}(F_\lambda(\varphi(u)))'_\varphi)^{-5}\{\{\varphi(u)[1 - 3e^{2iu}(F_\lambda(\varphi(u)))'_\varphi(F_\lambda(\varphi(u)))''_\varphi\varphi(u)\\
&+e^{iu}(F_\lambda(\varphi(u)))'''_\varphi\varphi^2(u) + 2e^{3iu}((F_\lambda(\varphi(u)))'_\varphi)^3 + 2e^{iu}(F_\lambda(\varphi(u)))'_\varphi\\
&-e^{2iu}(F_\lambda(\varphi(u)))'''_\varphi(F_\lambda(\varphi(u)))'_\varphi\varphi^2(u) + 3e^{2iu}((F_\lambda(\varphi(u)))''_\varphi\varphi(u))^2\\
&-6e^{2iu}((F_\lambda(\varphi(u)))'_\varphi)^2 + 6e^{iu}(F_\lambda(\varphi(u)))''_\varphi\varphi(u) + e^{4iu}((F_\lambda(\varphi(u)))'_\varphi)^4\\
&-3e^{3iu}(F_\lambda(\varphi(u)))''_\varphi((F_\lambda(\varphi(u)))'_\varphi)^2\varphi(u)] - (r+1)^3q_r(\lambda)\\
&\times\exp\{iu(r+1)\}(1 - e^{iu}(F_\lambda(\varphi(u)))'_\varphi)^5\}(\varphi(u) - q_r(\lambda)\\
&\times\exp\{iu(r+1)\})^2 - 3(\varphi(u) - q_r(\lambda)\exp\{iu(r+1)\})(\varphi(u)\\
&-q_r(\lambda)e^{iur}(1 - e^{iu}(F_\lambda(\varphi(u)))'_\varphi))[\varphi(u)(1 - e^{2iu}((F_\lambda(\varphi(u)))'_\varphi)^2\\
&+e^{iu}(F_\lambda(\varphi(u)))''_\varphi\varphi(u)) - (r+1)^2q_r(\lambda)e^{iur}(1\\
&-e^{iu}(F_\lambda(\varphi(u)))'_\varphi)^3](1 - e^{iu}(F_\lambda(\varphi(u)))'_\varphi) + 2(\varphi(u) - (r+1)q_r(\lambda)\\
&\times\exp\{iu(r+1)\}(1 - e^{iu}(F_\lambda(\varphi(u)))'_\varphi))^3(1 - e^{iu}(F_\lambda(\varphi(u)))'_\varphi)^2\}.
\end{aligned}
\tag{3}
$$

Firstly we will prove that

$$\sigma_r\sqrt{S} \to \infty \tag{4}$$

uniformly in r. For this we consider the different values of r, j in three domains of variation of N, n: $n/N \to 0$, $0 < C_1 \leqslant n/N \leqslant C_2 < \infty$, $n/N \to \infty$.

Let $n/N \to 0$. Using (1.2) and (1.8) it is not hard to get

$$B_\lambda = \frac{\lambda^2F''(\lambda)}{F(\lambda)} + \frac{\lambda F'(\lambda)}{F(\lambda)} - \left(\frac{\lambda F'(\lambda)}{F(\lambda)}\right)^2. \tag{5}$$

Let $r \geqslant 2, r \neq j$. Using (1.5) we find that the equality $q_r(\lambda) = 0$ is valid as $r < j$ and $q_r(\lambda) = O(\lambda^2)$ as $r > j$. Therefore from (2), (5), (1.2), (1.5), (1.7) and the condition $S = N(1 - q_r(\lambda))(1 + o(1))$ we obtain $\sigma_r^2 \geqslant C_3\lambda^j$, $S \geqslant C_4N$, where the symbols $C_3, C_4, \ldots$ here and below denote positive constants. Hence, using the relation $N\lambda^j \to \infty$, we get (4). It is clear that these reasons are true uniformly in r, including the case $r \to \infty$.

Let $r = 1, j > 1$. Then from (2), (1.5) it follows that $q_r(\lambda) = 0$ and $S\sigma_1^2 = Nj^2(p_j/p_0)\lambda^j(1 + o(1))$. This and the condition $N\lambda^j \to \infty$ imply (4).

Let $r = j \geqslant 2$. Then again using (2), (5), (1.2), (1.5) and (1.7) we find that $q_r(\lambda) = (p_j/p_0)\lambda^j + O(\lambda^{2j})$ and $\sigma_j^2 \geqslant C_5(\lambda^{2j} + \lambda^{j+s})$; therefore from the relation $S = N(1 - q_r(\lambda))(1 + o(1))$ and the condition $N\lambda^\omega \to \infty$ we obtain (4). If $r = 0$, $j \geqslant 3$ then by analogy we find that $S\sigma_0^2 \geqslant C_6(\lambda^{2j} + \lambda^{j+s}) \to \infty$. If $r = 0$, $j = 1$ then $S\sigma_0^2 \geqslant C_7N\lambda^2 \to \infty$ because in this case $\omega = 2$. If $r = 0$, $j = 2$, $p_3 > 0$ then $S\sigma_0^2 \geqslant C_8N\lambda^3$ and we have (4). In the case $r = 0$, $j = 2$, $p_3 = p_4 = 0$ we see that $S\sigma_0^2 \geqslant C_9N\lambda^4$ and relation $N\lambda^4 \to \infty$ implies (4). If $r = 1$, $j = 1$ then $S\sigma_1^2 \geqslant C_{10}N\lambda^2 \to \infty$.

Thus we have proved (4) in the case $n/N \to 0$. Now consider the case $0 < C_1 \leqslant n/N \leqslant C_2 < \infty$. From Lemma 2.2.1 it follows that $0 < C_{12} \leqslant \lambda \leqslant C_{13} < 1$. As we see from (2) and (1.5) the variance σ_r^2 and the probability $q_r(\lambda)$ continuously depend on λ. This means that $\sigma_r^2 \geqslant C_{14} > 0$ and $S \geqslant C_{15}N$; therefore (4) is true for any fixed r. If $r \to \infty$, then from (1.5) and relations $C_4 < 1$, $q_r(\lambda) \to 0$, $r^2 q_r(\lambda) \to 0$ we find (4) again.

Finally let $n/N \to \infty$, $n/N^2 \to 0$. By Lemma 2.2.1 $\lambda \to 1$. Then $\lambda \in [C_{12}, 1]$ and we can repeat the preceding arguments for fixed r. Let $r \to \infty$. Expanding $F(\lambda)$ and $F'(\lambda)$ into a series in the neighbourhood of $\lambda = 1$ from (1.4) and (1.5) we obtain

$$\lambda = 1 - B(N/n) + O((N/n)^2). \tag{6}$$

By Lemma 1.3.12 and (2.2.4)

$$q_r(\lambda) = O(r^{-3/2}(\lambda/F(\lambda))^r); \tag{7}$$

therefore, from (6) we get

$$r^2 q_r(\lambda) \leqslant C_{16} r^{1/2} \exp\{-C_{17} Nr/n\}. \tag{8}$$

It is not hard to obtain from (2.2.1), (1.7) and (6) that

$$\sigma^2 = O((n/N)^3). \tag{9}$$

By (8) and (9)

$$r^2 q_r(\lambda)/\sigma^2 \to 0; \tag{10}$$

therefore from (2), (9) and the inequality $r \leqslant n$ we see that (4) is true. It is clear for all these reasons that the relation (4) holds uniformly in r.

Since

$$\varphi_S^{(r)}(u) = \exp\left\{-\frac{iSa_r u}{\sigma_r\sqrt{S}}\right\}\left(\varphi^{(r)}\left(\frac{u}{\sigma_r\sqrt{S}}\right)\right)^S,$$

it follows that

$$\ln\varphi_S^{(r)}(u) = -\frac{iSa_r u}{\sigma_r\sqrt{S}} + S\ln\varphi^{(r)}\left(\frac{u}{\sigma_r\sqrt{S}}\right). \tag{11}$$

For sufficiently small u

$$\begin{aligned}\ln\varphi^{(r)}(u) = u\left(\frac{\partial\ln\varphi^{(r)}(u)}{\partial u}\right)_{u=0} + \frac{u^2}{2}\left(\frac{\partial^2\ln\varphi^{(r)}(u)}{\partial^2 u}\right)_{u=0} \\ + \frac{u^3}{3!}Q(u) = iua_r - \frac{u^2\sigma_r^2}{2} + \frac{u^3}{6}Q(u),\end{aligned} \tag{12}$$

where

$$|Q(u)| \leqslant 2\max_{|\tau|\leqslant|u|}\left|(\ln\varphi^{(r)}(u))_\tau'''\right|. \tag{13}$$

From (4) it follows that for any fixed u the relation $u/(\sigma\sqrt{S}) \to 0$ holds. Therefore using (12) and (13) we obtain

$$\ln\varphi_S^{(r)}(u) = -\frac{u^2}{2} + \frac{u^3}{6\sigma_r^3\sqrt{S}}Q\left(\frac{u}{\sigma_r\sqrt{S}}\right). \tag{14}$$

To prove Lemma 1 it suffices to establish that the second term on the right-hand side of (14) tends to zero.

Using (13) we obtain

$$\left|\frac{u^3}{6\sigma_r^3\sqrt{S}}Q\left(\frac{u}{\sigma_r\sqrt{S}}\right)\right| \leqslant \frac{|u|^3}{3}Q_1(u), \tag{15}$$

where

$$Q_1(u) = \max_{|\tau|\leqslant|u/(\sigma_r\sqrt{S})|}\left|(\ln\varphi^{(r)}(\tau))_\tau'''\right|/(\sigma_r^3\sqrt{S}). \tag{16}$$

Consider the case $n/N \to \infty$, $n/N^2 \to 0$. For sufficiently large N, n and sufficiently small u

$$|\varphi(u) - q_r(\lambda)\exp\{iu(r+1)\}| \geqslant |\varphi(u)| - q_r(\lambda) \geqslant C_{18} > 0. \tag{17}$$

The last inequality is true because there exists the value of $\nu^{(1)}$ which differs from $r+1$; hence $q_r(\lambda) \to q_r(1) \leqslant C_{19} < 1$ and we can make $|\varphi(u)|$ arbitrarily close to one by choosing a sufficiently small u. It is easy to see that $|e^{iu}(F_\lambda(\varphi(u)))_\varphi'| \leqslant m$; therefore by (1.4) and (1.8)

$$|1 - e^{iu}(F_\lambda(\varphi(u)))_\varphi'| \geqslant 1 - m = N/(N+n). \tag{18}$$

Using (3), (17) and (18) to estimate $(\ln\varphi^{(r)}(u))_u'''$ and taking into account the limitedness of $F'''(1)$ and feasibility of the condition $r \to \infty$ we obtain that for sufficiently large N, n and sufficiently small u

$$\left|(\ln\varphi^{(r)}(u))_u'''\right| \leqslant C_{20}((n/N)^5 + r^3 q_r(\lambda)). \tag{19}$$

From (8) we find that

$$r^3 q_r(\lambda) \leqslant C_{16} r^{3/2}\exp\{-C_{17}Nr/n\}. \tag{20}$$

This together with the inequality $r \leqslant n$ and the relation $r \to \infty$ implies that $r^3 q_r(\lambda) = o((n/N)^5)$. Therefore from (19) it follows that

$$\left|(\ln\varphi^{(r)}(u))_u'''\right| \leqslant C_{21}(n/N)^5. \tag{21}$$

It is not hard to obtain from (2.3.2), (2.2.1) and (6) that

$$\sigma^2 = O((n/N)^3). \tag{22}$$

Using (8) and (22) we have that, for $r \to \infty$, the relation $r^2 q_r(\lambda)/\sigma^2 \to 0$ is valid. Therefore (2) and (22) for any r imply that

$$\sigma_r^2 \geqslant C_{22}(n/N)^3.$$

From this and (16), (21) we find that $Q_1(u) \leqslant C_{23}\sqrt{n}/N$; hence $Q_1(u) \to 0$. This together with (14) and (15) implies the assertion of Lemma 1 in this case.

Let $N, n \to \infty$ in such a way that $n/N \leqslant C_2 < \infty$. We will prove that for sufficiently large N, n and sufficiently small u

$$\left|(\ln\varphi^{(r)}(u))_u'''\right|/\sigma_r^2 \leqslant C_{24}. \tag{23}$$

To this end we divide the following proof into four parts.

1. Let $0 < C_1 \leqslant n/N \leqslant C_2 < \infty$. In this case it follows from (1.4), (1.8) and Lemma 1.2.1 that $0 < C_{12} \leqslant \lambda \leqslant C_{13} < 1$, $0 < C_{25} \leqslant m \leqslant C_{26} < 1$. As we saw above, $q_r(\lambda) \leqslant C_{27} < 1$; therefore the inequality (17) is true for sufficiently large

N, n and sufficiently small u. It is clear also that $|e^{iu}(F_\lambda(\varphi(u)))'_\varphi| \leqslant m \leqslant C_{26} < 1$; therefore

$$|1 - e^{iu}(F_\lambda(\varphi(u)))'_\varphi| \geqslant C_{28} > 0. \tag{24}$$

In the proof of Lemma 2.2.3 we saw that $\lambda/F(\lambda) \leqslant C_{29} < 1$; therefore from (7) it follows that $r^3 q_r(\lambda) \to 0$ for $r \to \infty$. This and (3), (17) and (24) imply that

$$\left|(\ln \varphi^{(r)}(u))'''_u\right| \leqslant C_{30}.$$

Using (1.5) and (2) we obtain that $\sigma_r^2 \geqslant C_{31}$; therefore the inequality (23) holds.

2. Let $N, n \to \infty$ in such a way that $n/N \to 0$, $N\lambda^j \to \infty$, $r \geqslant 2$. By Lemma 1.2.1 and (1.8) $\lambda \to 0$, $m \to 0$; therefore (1.5), (17) and (18) give us

$$|\varphi(u) - q_r(\lambda)\exp\{iu(r+1)\}| \geqslant C_{32}, \quad |1 - e^{iu}(F_\lambda(\varphi(u)))'_\varphi| \geqslant C_{33}. \tag{25}$$

From this and Lemma 1.2.1 we obtain

$$\left|(\ln \varphi^{(r)}(u))'''_u\right| \leqslant C_{34}\lambda \leqslant C_{35}(n/N)^j. \tag{26}$$

Using (1.6), (2) and (2.3.4) we get that $\sigma_r^2 \geqslant C_{36}(n/N)^j$ and (23) follows from (26).

3. Let $N, n \to \infty$, $n/N \to 0$, $r = 1$. If $j = 1$ then, as we saw above, $\sigma_r^2 \geqslant C_{37}\lambda^2$. If $j > 1$ then by analogy $\sigma_1^2 \geqslant C_{38}\lambda^j$. Using (3) and (25) it is not hard to get that $|(\ln \varphi^{(r)}(u))'''_u| \leqslant C_{39}\lambda^2$ for $j = 1$ and $|(\ln \varphi^{(r)}(u))'''_u| \leqslant C_{40}\lambda^j$ for $j > 1$. These relations imply (23).

4. Let $N, n \to \infty$, $n/N \to 0$, $r = 0$. Using (2.3.1) and (2.1.4) we can easily deduce

$$|\varphi(u) - q_0(\lambda)e^{iu}| \geqslant C_{41}\lambda^j. \tag{27}$$

Using (25), (3) and making the necessary calculations we find that for $j = 1$

$$\left|(\ln \varphi^{(0)}(u))'''_u\right| \leqslant C_{42}\lambda. \tag{28}$$

From (2), (5), (1.2) and (1.7) it is easy to get that $\sigma_0^2 \geqslant C_{36}\lambda$; therefore (28) implies (23). If $j = 2$, then $|(\ln \varphi^{(0)}(u))'''_u| \leqslant C_{42}\lambda$, $\sigma_0^2 \geqslant C_{43}\lambda$ for $p_3 > 0$ and $|(\ln \varphi^{(0)}(u))'''_u| \leqslant C_{44}$, $\sigma_0^2 \geqslant C_{45}\lambda^2$ for $p_3 = 0$. Thus for $j = 2$, (23) is true. Let $j \geqslant 3$. As above we see that $|(\ln \varphi^{(0)}(u))'''_u| \leqslant C_{46}\lambda^j$, $\sigma_0^2 \geqslant C_{47}\lambda^j$ for $s = 0$ and $|(\ln \varphi^{(0)}(u))'''_u| \leqslant C_{48}\lambda^{\omega - j}$, $\sigma_0^2 \geqslant C_{49}\lambda^{\omega - j}$ for $s > 0$. The relation (23) is obtained from these inequalities.

To conclude the proof it remains to note that by (4), (14) and (23)

$$\ln \varphi_S^{(r)}(u) \to -u^2/2.$$

Lemma 2. *Let $N, n \to \infty$ in such a way that $n/N^2 \to 0$, $n/N \geqslant C > 0$, $F'''(1) < \infty$ if $n/N \to \infty$. Then for $S = N(1 - q_r(\lambda))(1 + o(1))$ and non-negative integers h divisible by d*

$$\mathbf{P}\{\zeta_S^{(r)} = N + h\} = d(\sigma_r\sqrt{2\pi S})^{-1}\exp\{-u_r^2/2\}(1 + o(1))$$

uniformly in $u_r = (N + h - Sa_r)/(\sigma_r\sqrt{S})$ lying in any finite interval. This assertion is valid for $r \to \infty$ as well.

Proof. We follow the scheme of proving Lemma 2.3.2. We represent the probability $\mathbf{P}\{\zeta_S^{(r)} = N + h\}$ as the integral

$$\mathbf{P}\{\zeta_S^{(r)} = N + h\} = \frac{d}{2\pi\sigma_r\sqrt{S}} \int\limits_{-d^{-1}\pi\sigma_r\sqrt{S}}^{d^{-1}\pi\sigma_r\sqrt{S}} e^{-izu}\varphi_S^{(r)}(u)du, \tag{29}$$

where $z = (N + h - Sa_r)/(\sigma_r\sqrt{S})$. Using (2.3.15) we can divide the difference

$$R_S = 2\pi[d^{-1}\sigma_r\sqrt{S}\,\mathbf{P}\{\zeta_S^{(r)} = N + h\} - (2\pi)^{-1/2}e^{-z^2/2}]$$

into the sum of four integrals: $R_S = I_1 + I_2 + I_3 + I_4$, where

$$\begin{aligned} I_1 &= \int\limits_{-A}^{A} e^{-izu}[\varphi_S^{(r)}(u) - e^{-u^2/2}]du, \\ I_2 &= \int\limits_{A<|u|\leqslant\varepsilon\sigma_r\sqrt{S}} e^{-izu}\varphi_S^{(r)}(u)du, \\ I_3 &= \int\limits_{\varepsilon\sigma_r\sqrt{S}<|u|\leqslant d^{-1}\pi\sigma_r\sqrt{S}} e^{-izu}\varphi_S^{(r)}(u)du, \\ I_4 &= -\int\limits_{A<|u|} \exp\{-izu - u^2/2\}du; \end{aligned} \tag{30}$$

the positive constants A and ε will be chosen later.

Using Lemma 2.2.1 from Lemma 1 we obtain $I_1 \to 0$. To estimate I_4 we can use the inequality (2.3.17).

To estimate the integrals I_2 and I_3 we first consider the case $0 < C_1 \leqslant n/N \leqslant C_2 < \infty$. Using (23), (12) and (13) we obtain that for sufficiently small u

$$|\varphi^{(r)}(u)| \leqslant \exp\{-C_3u^2\sigma_r^2\}; \tag{31}$$

hereafter $C_3, C_4, \ldots$ denote some positive constants. Using the relation

$$\varphi_S^{(r)}(u) = \exp\{-Sa_ru/(\sigma_r\sqrt{S})\}(\varphi^{(r)}(u/(\sigma_r\sqrt{S})))^S, \tag{32}$$

from (30) and (31) we get an estimate similar to (2.3.18); hence the integral I_2 can be made arbitrarily small by choosing N, n, A.

To estimate I_3 we note that if $\varepsilon < |u| \leqslant \pi/d$, then

$$|\varphi^{(r)}(u)| \leqslant e^{-C_4}. \tag{33}$$

It is not hard to see from (2) that $0 < C_5 \leqslant \sigma_r \leqslant C_6 < \infty$; therefore by (30) and (33)

$$|I_3| \leqslant 2\pi d^{-1}\sigma_r\sqrt{S}e^{-C_4S} \leqslant C_{10}\sqrt{S}e^{-C_4S} \to 0. \tag{34}$$

Let $n/N \to \infty$, $n/N^2 \to 0$. It is clear that in this case the inequality (33) is true. By analogy with the proof of Lemma 1 from (1.5) and (2) it follows that

$$\sigma_r^2 = O((n/N)^3). \tag{35}$$

This together with (33) and relations $S = N(1 - q_r(\lambda))(1 + o(1))$, $n = o(N^2)$ implies that

$$|I_3| \leqslant 2\pi d^{-1}\sigma_r\sqrt{S}e^{-C_4 S} \leqslant C_{11}N^2 e^{-C_{12}N} \to 0.$$

To estimate the integral I_2 we represent it as a sum of two integrals: $I_2 = I_2' + I_2''$ whose integration domains are the sets $\{u: \ A < |u| \leqslant \varepsilon_1 N/\sqrt{n}\}$, $\{u: \ \varepsilon_1 N/\sqrt{n} < |u| \leqslant \varepsilon\sigma_r\sqrt{S}\}$. Using (6), (14)–(16) and (19) we obtain that in the first domain for sufficiently small $\varepsilon_1 > 0$

$$\left|\varphi_S^{(r)}(u)\right| \leqslant \exp\{-C_{13}u^2\}$$

and find the estimation (2.3.36). Therefore I_2' can be made arbitrarily small by choosing A.

To estimate I_2'' we will prove that if u is sufficiently small and

$$|u|(r+1)^2 \leqslant 2, \tag{36}$$

then there exists such positive constant C_{14} that

$$\left|\varphi^{(r)}(u)\right| = \left|\frac{\varphi(u) - q_r(\lambda)\exp\{iu(r+1)\}}{1 - q_r(\lambda)}\right| \leqslant |\varphi(u)|^{C_{14}}. \tag{37}$$

By Lemma 1.3.3 as $u \to 0$

$$\begin{aligned} \varphi(u) &= \exp\{-\sqrt{-2iu/B_\lambda} + o(\sqrt{u})\} \\ &= \exp\{-\sqrt{|u|/B_\lambda}(1 - iu/|u|) + o(\sqrt{u})\}. \end{aligned} \tag{38}$$

This means that

$$\varphi(u) = R(\cos\alpha + i\sin\alpha), \tag{39}$$

where

$$R = |\varphi(u)| = \exp\{-\sqrt{|u|/B_\lambda} + o(\sqrt{u})\}, \quad \alpha = u/\sqrt{B_\lambda|u|} + o(\sqrt{u}).$$

Using (36) we can find that for sufficiently small u

$$\begin{aligned} \cos\alpha &\geqslant 1 - \alpha^2/2 \geqslant 1 - |u|/B_\lambda, \\ \cos(u(r+1)) &\geqslant 1 - (u(r+1))^2/2 \geqslant 1 - |u|; \end{aligned}$$

therefore $\cos\alpha\cos(u(r+1)) \geqslant 1 - |u|(1 + B_\lambda^{-1})$. It is clear that $R \leqslant \exp\{-\sqrt{|u|}/(2\sqrt{B_\lambda})\}$ and there exists such small $\delta = \delta(u)$ that

$$\begin{aligned} \cos\alpha\cos(u(r+1)) &\geqslant 1 - |u|(1 + B_\lambda^{-1}) \\ &\geqslant \exp\{-\delta\sqrt{|u|}/(2\sqrt{B_\lambda})\} \geqslant R^\delta. \end{aligned} \tag{40}$$

Using (36) it is not hard to get that for sufficiently small u the inequality $\sin\alpha\sin(u(r+1)) \geqslant 0$ holds; therefore from (40) we obtain

$$\cos\alpha\cos(u(r+1)) + \sin\alpha\sin(u(r+1)) \geqslant R^\delta;$$

hence

$$\begin{aligned} |\varphi(u) - q_r(\lambda)\exp\{iu(r+1)\}|^2 &= (R\cos\alpha - q_r(\lambda)\cos(u(r+1)))^2 \\ +(R\sin\alpha - q_r(\lambda)\sin(u(r+1)))^2 &= R^2 + q_r^2(\lambda) - 2Rq_r(\lambda) \\ \times(\cos\alpha\cos(u(r+1)) + \sin\alpha\sin(u(r+1))) &\leqslant R^2 + q_r^2(\lambda) - 2R^{1+\delta}q_r(\lambda). \end{aligned} \tag{41}$$

Using (1.5) we can estimate $q_r(\lambda)$. Taking into account that the quantity $R = |\varphi(u)|$ can be made arbitrarily close to one by choosing u we find that there exists a constant C_{14} such that

$$R^2 + q_r^2(\lambda) - 2R^{1+\delta}q_r(\lambda) - (1 - q_r(\lambda))^2 R^{2C_{14}} \leqslant 0.$$

This together with (41) implies (37).

If r is fixed then the inequality (36) is true for sufficiently small u therefore we can use (37). From (32), (37) and (38) it follows that in the integration domain of I_2''

$$|\varphi^{(r)}(u/(\sigma_r\sqrt{S}))| \leqslant \exp\{-C_{15}\sqrt{|u|/(\sigma_r\sqrt{S})}\}. \tag{42}$$

From this and (35) we obtain by analogy with (2.3.38)

$$|I_2''| \leqslant 2 \int\limits_{\varepsilon_1 N/\sqrt{n}}^{\infty} \exp\{-C_{15}\sqrt{u}(N^2/n)^{3/4}\}du. \tag{43}$$

The last expression tends to zero by the relation $N^2/n \to \infty$. Let $r \to \infty$. We represent the integral I_2'' as the sum $I_2'' = I_2''(1) + I_2''(2)$, where the integration domains of $I_2''(1)$ and $I_2''(2)$ are

$$S_1 = \{u : \varepsilon_1 N/\sqrt{n} \leqslant |u| < 2\sigma_r\sqrt{S}/(r+1)^2\},$$
$$S_2 = \{u : 2\sigma_r\sqrt{S}/(r+1)^2 < |u| \leqslant \varepsilon\sigma_r\sqrt{S}\}.$$

Note that S_2 can be empty. In S_1 the inequality (36) is valid; therefore we can use the estimate (37) and get the relation similar to (43):

$$|I_2''(1)| \leqslant 2 \int\limits_{\varepsilon_1 N/\sqrt{n}}^{\infty} \exp\{-C_{15}\sqrt{u}(N^2/n)^{3/4}\}du \to 0. \tag{44}$$

In the domain S_2 we can see from (2.3.37) that

$$|\varphi(u/(\sigma_r\sqrt{S}))| \leqslant \exp\{-C_{16}\sqrt{|u|/(\sigma_r\sqrt{S})}\}. \tag{45}$$

Using (20) we obtain from (1) and (45)

$$\left|\varphi^{(r)}\left(\frac{u}{\sigma_r\sqrt{S}}\right)\right| \leqslant \exp\left\{S\left(-C_{16}\sqrt{\frac{|u|}{\sigma_r\sqrt{S}}} + \frac{C_{17}}{r^{3/2}}\right)\right\}. \tag{46}$$

It is not hard to see that for $r \to \infty$ and $|u| > 2\sigma_r\sqrt{S}/(r+1)^2$ the relation $r^{-3/2} = o((|u|/(\sigma_r\sqrt{S}))^{1/2})$ is valid; therefore from (46) it follows that in S_2 the estimation (42) is true. Hence, for $I_2''(2)$ there exists the inequality similar to (44). This completes the proof of Lemma 2.

Lemma 3. *Let $N, n \to \infty$ in such a way that $n/N \to 0$, n be divided by d and $N^{\omega(r)} \to \infty$, where*

$$\begin{aligned}
&\omega(0) = 2j + l; && \\
&\omega(1) = 3 && \text{for } j = 1; \\
&\omega(1) = j + l && \text{for } j > 1; \\
&\omega(r) = j + l && \text{for } r \geqslant 2, r \neq j; \\
&\omega(r) = \max\{\omega, j + l\} && \text{for } r = j \geqslant 2.
\end{aligned}$$

Then for $S = N(1 - q_r(\lambda))(1 + o(1))$ the assertion of Lemma 2 is valid.

Proof. Lemma 3 is proved by analogy with Lemma 2. For this we represent the probability $\mathbf{P}\{\zeta_S^{(r)} = N + h\}$ as the integral (29) and the difference R_S as the sum of four integrals (30). As above it is easy to see that the estimations of I_1 and I_4 are valid.

Note that the inequality (31) is true; therefore from (31) and (32) follows the estimation similar to (2.3.18). This means that the integral I_2 is arbitrarily small at sufficiently large N, n, A. Thus we need to consider only the integral I_3.

If $r \geqslant 2$, then from (1), (1.4) and (1.5) we obtain that

$$|\varphi^{(r)}(u)| = |\varphi(u)|(1 + O((n/N)^2)). \tag{47}$$

Let $j = 1$. Using Lemma 2.2.1 and the relations (2), (1.2), (1.5), (1.7), (2.3.4) we can get $\sigma_r^2 = O(n/N)$. From this by analogy with (2.3.23) we obtain

$$|\varphi(u/(\sigma_r\sqrt{S}))| \leqslant \exp\{-C_1 n/N\};$$

the meaning of the denotions $C_1, C_2, \ldots$ is preserved. This together with (30), (32), (47), (1.4) implies that

$$|I_3| \leqslant C_2\sqrt{n}e^{-C_3 n} \to 0.$$

If $j > 1$ but $l = 0$, we can get again the estimations similar to (2.3.24) and (2.3.25). To estimate $|\varphi(u)|$ in the case $j > 1$, $l > 0$, we represent the integral I_3 as the sum $I_3' + I_3''$ and use the relations (2.3.27), (2.3.28), (2.3.30) and (2.3.31).

If $r = 1, j > 1$, then (1), (1.1), (1.2), (1.4), (1.5) and (1.8) imply that for the characteristic function $\varphi^{(1)}(u)$ the inequalities (2.3.27) and (2.3.30) are true and the estimate of I_3 is obtained in Lemma 2.3.2. If $j = 1$ then using (1), (1.1), (1.2), (1.4), (1.5), (1.8) we get

$$|\varphi^{(1)}(u)| \leqslant \exp\{-C_4\lambda^2(1 - \cos(2u)) - C_5\lambda^3(1 - \cos(3u))\}. \tag{48}$$

Representing I_3 as the sum $I_3' + I_3''$ by analogy with Lemma 2.3.2 we find that $I_3' \to 0$. To estimate I_3'' we expand $1 - \cos(2u)$ into a series in the neighbourhood of $u = \pi$ and obtain from (48)

$$|\varphi^{(1)}(u)| \leqslant \exp\{-C_6\lambda^3 - C_7\lambda^2(u - \pi)^2\}.$$

From this it follows that

$$|I_3''| \leqslant \int\limits_{(\pi-\varepsilon)\sigma_1\sqrt{S}<|u|\leqslant\pi\sigma_1\sqrt{S}} |\varphi^{(1)}(u/(\sigma_1\sqrt{S}))|^S du$$

$$= \sigma_1\sqrt{S}\int\limits_{-\varepsilon}^{\varepsilon} |\varphi^{(1)}(\pi+u)|du;$$

therefore

$$|I_3''| \leqslant \sigma_1\sqrt{S}e^{-C_8N\lambda^3}\int\limits_{-\varepsilon}^{\varepsilon} e^{-C_9\lambda^2u^2S}du$$

$$\leqslant C_{10}\lambda\sqrt{N}e^{-C_8N\lambda^3}\int\limits_{-\varepsilon}^{\varepsilon} e^{-C_9\lambda^2u^2S}du \leqslant C_{11}e^{-C_8N\lambda^3}\int\limits_{-\infty}^{\infty} e^{-C_{12}u^2}du.$$

The last expression tends to zero by the condition $N\lambda^3 \to \infty$.

Finally, let $r = 0$. If $j = 1$ then from (1) we deduce that $|\varphi^{(0)}(u)| \leqslant e^{-C_{13}\lambda}$; therefore from the relations $\sigma_0^2 = O(\lambda)$, $1 - q_0(\lambda) = O(\lambda)$ and $N\lambda^2 \to \infty$ it follows that

$$|I_3| \leqslant C_{14}\sqrt{N\lambda^2}\exp\{-C_{15}N\lambda^2\} \to 0.$$

It is easy to see also that in the case $j > 1, l = 0$

$$|I_3| \leqslant C_{14}\sqrt{N\lambda^{2j}}\exp\{-C_{15}N\lambda^{2j}\} \to 0.$$

If $j > 1, l > 0$ then for the characteristic function $\varphi^{(0)}(u)$ the estimations similar to (2.3.27) and (2.3.30) are true. Therefore as in Lemma 2.3.2 we can represent I_3 as $I_3' + I_3''$. In distinction from the proof of Lemma 2.3.2 in this case we must take into consideration that by (1.5)

$$S = N(1 - q_r(\lambda))(1 + o(1)) = O(N\lambda^j)$$

and for the convergence of I_3 to zero it is necessary to have $N\lambda^{2j+l} \to 0$.

3. The limit behaviour of the sum of auxiliary random variables in the critical case

Let us consider the sum $\zeta_S^{(r)} = \nu_{(r)}^{(1)} + \cdots + \nu_{(r)}^{(S)}$ in the critical case $\lambda = 1$ where λ is the parameter of the distribution (1.1). Below we will prove Lemma 1 on the local convergence of the distribution $\zeta_S^{(r)}$ to the limit law and Lemma 2 on large deviations at the same convergence.

Lemma 1. *Let $N \to \infty$, $S = N(1-q_r(1))(1+o(1))$, $z = (N+h)BN^{-2}$, where h is a natural number divisible by d, B is the variance of the distribution* (1.1), $\lambda = 1$. *Then*

$$B^{-1}N^2\mathbf{P}\left\{B\zeta_S^{(r)}/N^2 = z\right\} = \frac{d(1+o(1))}{\sqrt{2\pi}}z^{-3/2}e^{-1/(2z)}$$

uniformly in z, $0 < z_0 \leqslant z \leqslant z_1 < \infty$. This equality is valid for $r \to \infty$ as well.

Proof. Let $\varphi_{S,r}(u)$ be the characteristic function of the variable $B\zeta_S^{(r)}/N^2$. From (2.38) and the equality

$$\varphi_{S,r}(u) = \left(\frac{\varphi(uB/N^2) - q_r(1)\exp\{iu(r+1)B/N^2\}}{1-q_r(1)}\right)^S \tag{1}$$

it can easily be derived that as $r = o(N)$

$$\varphi_{S,r}(u) \to \exp\{-\sqrt{-2iu}\} = \exp\{-\sqrt{|u|}(1 - iu/|u|)\} \tag{2}$$

uniformly in u lying in any finite interval. If $r \geqslant C_1 N$, and here and below $C_1, C_2, \dots$ are some positive constants, then by Lemma 1.3.12 $Nq_r(1) = o(1)$, therefore (1) implies (2). As in the proof of Lemma 2.3.3 the assertion of Lemma 1 for fixed r immediately follows from (2) and Theorem 1.3.6. If $r \to \infty$ then the sums $\zeta_S^{(r)}$ form an array scheme. Therefore we cannot use Theorem 1.3.6 and need to prove the assertion of Lemma 1 in this case.

By the inversion formula, the probability

$$\mathbf{P}\left\{\zeta_S^{(r)} = N + h\right\} = \mathbf{P}\left\{B\zeta_S^{(r)}/N^2 = z\right\}$$

can be represented in the form

$$\mathbf{P}\left\{\frac{B\zeta_S^{(r)}}{N^2} = z\right\} = \frac{dB}{2\pi N^2}\int_{-\pi N^2/(dB)}^{\pi N^2/(dB)} e^{-izu}\left(\varphi^{(r)}\left(\frac{uB}{N^2}\right)\right)^S du.$$

Since

$$(2\pi z^3 e^{1/z})^{-1/2} = (2\pi)^{-1}\int_{-\infty}^{\infty} \exp\{-izu - \sqrt{|u|}(1 - iu/|u|)\}du,$$

the difference $R_S = 2\pi N^2(dB)^{-1}\,\mathbf{P}\left\{B\zeta_S^{(r)}/N^2 = z\right\} - (2\pi)^{1/2}(z^3e^{1/z})^{-1/2}$ can be represented as the sum $R_S = I_1 + I_2 + I_3 + I_4$, where

$$I_1 = \int_{-A}^{A} e^{-izu}[(\varphi^{(r)}(uB/N^2))^S - \exp\{-\sqrt{|u|}(1 - iu/|u|)\}]du,$$

$$I_2 = \int_{A<|u|\leqslant \varepsilon N^2} e^{-izu}(\varphi^{(r)}(uB/N^2))^S du,$$

$$I_3 = \int_{\varepsilon N^2<|u|\leqslant \pi N^2/(dB)} e^{-izu}(\varphi^{(r)}(uB/N^2))^S du, \tag{3}$$

$$I_4 = -\int_{A<|u|} \exp\{-izu - \sqrt{|u|}(1 - iu/|u|)\}du;$$

the positive constants A and ε will be chosen later.

Since $\varphi_{S,r}(u) = (\varphi^{(r)}(uB/N^2))^S$, it follows from (2) and (3) that $I_1 \to 0$. It is clear also that the quantity $|I_4|$ can be made arbitrarily small by choosing A.

Let us estimate I_3. For $\varepsilon < |u| \leqslant \pi/d$, the inequality $|\varphi(u)| \leqslant e^{-C_2}$ is valid. From this and (2.1) it follows that for sufficiently large N

$$|\varphi^{(r)}(u)| \leqslant \frac{|\varphi(u)| + q_r(1)}{1 - q_r(1)} \leqslant e^{-C_2}\frac{1 + e^{C_2}q_r(1)}{1 - q_r(1)}.$$

Since $q_r(1) \to 0$ as $r \to \infty$ we obtain that for sufficiently large r $|\varphi^{(r)}(u)| \leqslant e^{-C_2/2}$; therefore

$$|I_3| \leqslant 2\pi N^2 d^{-1} e^{-C_2 N/2} \to 0.$$

To estimate the integral I_2 we represent it as the sum $I_2 = I_2' + I_2''$, where the integration domains of I_2' and I_2'' are

$$S_2' = \{u : \ A < |u| \leqslant 2N^2/(B(r+1)^2)\},$$
$$S_2'' = \{u : \ 2N^2/(B(r+1)^2) < |u| \leqslant \varepsilon N^2\}.$$

If $2N^2/(B(r+1)^2) < A$ then the set S_2' is empty; therefore the left boundary of S_2'' is A.

It is easy to see that as $|u|(r+1)^2 \leqslant 2$, the proof of the inequality (2.37) is valid for $\lambda = 1$. From (2.37) and (2.38) we obtain that in S_2'

$$|\varphi^{(r)}(uB/N^2)| \leqslant \exp\{-C_3\sqrt{|u|}/N\}; \tag{4}$$

therefore

$$|I_2'| \leqslant \int_A^\infty \exp\{-C_4\sqrt{u}\}du \tag{5}$$

and I_2' can be made arbitrarily small by choosing A.

By (2.1)

$$\varphi^{(r)}(u) = \varphi(u)[1 - q_r(1)(1 - q_r(1))^{-1}(1 - \varphi^{-1}(u)\exp\{iu(r+1)\})];$$

hence, using again (2.38) and Lemma 1.3.12 we obtain that for sufficiently large N and small ε

$$\left|\varphi^{(r)}\left(\frac{uB}{N^2}\right)\right|^S \leqslant \exp\left\{S\left(-\frac{C_5\sqrt{|u|}}{N} + \frac{C_6}{r^{3/2}}\right)\right\}. \tag{6}$$

Since $|u| > 2N^2/(B(r+1))^2$, it follows from (6) that for $\varphi^{(r)}(uB/N^2)$ the estimate (4) is true in S_2''. Therefore for I_2'' the inequality similar to (5) is valid. Lemma 1 is proved.

Lemma 2. *Let $N \to \infty$, $S = N(1 - q_r(1))(1 + o(1))$, $z = B(N+n)N^{-2} \to \infty$, where n is a natural number divisible by d, $\lambda = 1$. Then*

$$B^{-1}N^2\,\mathbf{P}\left\{B\zeta_S^{(r)}/N^2 = z\right\} = d(2\pi)^{-1/2}z^{-3/2}(1 + o(1)).$$

This equality is valid for $r \to \infty$ as well.

Proof. By (2) the characteristic function $\varphi_{S,r}(u)$ converges to the characteristic function of the stable distribution with the exponent $\alpha = 1/2$ uniformly in u lying in any finite interval. This convergence is valid for any r including the case $r \to \infty$.

The limit distribution has the density $(2\pi x^3 e^{1/x})^{-1/2}$. Let the maximum span of the distribution $\nu_{(r)}^{(1)}$ be one. From Lemma 1.3.12 and (1.3) we obtain

$$\mathbf{P}\left\{\nu_{(r)}^{(1)} = k\right\} = O(k^{-3/2})$$

as $k \to \infty$. This means that for fixed r the conditions of Theorem 1.3.7 are valid and we get the assertion of Lemma 2.

We cannot use Theorem 1.3.7 if $d \neq 1$ or in the case of an array scheme as $r \to \infty$. That is why we prove below Lemma 2 by analogy with Lemma 2.5.2.

We represent the probability $\mathbf{P}\left\{\zeta_S^{(r)} = N + n\right\}$ as

$$\mathbf{P}\left\{\zeta_S^{(r)} = N + n\right\} = P_{S1}(n) + SP_{S2}(n) + P_{S3}(n), \tag{7}$$

where

$$P_{S1}(n) = \mathbf{P}\left\{\zeta_S^{(r)} = N + n;\ \nu_{(r)}^{(i)} \leqslant \gamma n,\ i = 1, \ldots, S\right\},$$

$$P_{S2}(n) = \mathbf{P}\left\{\zeta_S^{(r)} = N + n;\ \nu_{(r)}^{(i)} \leqslant \gamma n,\ i \leqslant S - 1,\ \nu_{(r)}^{(S)} > \gamma n\right\},$$

$$P_{S3}(n) = \mathbf{P}\left\{\zeta_S^{(r)} = N + n;\ \bigcup_{i \neq j} \left(\nu_{(r)}^{(i)} > \gamma n,\ \nu_{(r)}^{(j)} > \gamma n\right)\right\},$$

$$\gamma = (N^2/n)^{1/6}. \tag{8}$$

Let

$$R(\omega) = \sum_{k \leqslant \gamma(N+n)d^{-1}} \exp\{\omega(kd+1)\}\, \mathbf{P}\left\{\nu_{(r)}^{(1)} = kd + 1\right\}. \tag{9}$$

From (1.3) it follows that

$$\sum_{k \leqslant l} (kd+1)\, \mathbf{P}\left\{\nu_{(r)}^{(1)} = kd + 1\right\} \leqslant (1 - q_r(1))^{-1}$$

$$\times \sum_{k \leqslant l} (kd+1)\, \mathbf{P}\{\nu^{(1)} = kd + 1\}.$$

Therefore, from (2.5.14) we obtain that for sufiiciently large l

$$\sum_{k \leqslant l} (kd+1)\, \mathbf{P}\left\{\nu_{(r)}^{(1)} = kd + 1\right\} \leqslant C_1 \sqrt{l}; \tag{10}$$

hereafter $C_1, C_2, \ldots$ mean some positive constants. Similarly to (2.5.15) it is easy to get

$$\sum_{k > l} \mathbf{P}\left\{\nu_{(r)}^{(1)} = kd + 1\right\} \leqslant C_1 l^{-1/2}. \tag{11}$$

Using (9)–(11) we can deduce by analogy with (2.5.16) that

$$R(1/(\gamma n)) = 1 + o(N^{-1}). \tag{12}$$

Let $\nu^{(1)}(\gamma), \ldots, \nu^{(S)}(\gamma)$ be auxiliary independent identically distributed random variables such that

$$\mathbf{P}\{\nu^{(1)}(\gamma) = kd + 1\} = \exp\{(kd+1)/(\gamma(N+n))\}\, \mathbf{P}\left\{\nu_{(r)}^{(1)} = kd + 1\right\} / R(1/(\gamma n)),$$

where $k \leqslant \gamma(N+n)d^{-1}$. It is easy to see that

$$P_{S1}(n) = R^S(1/(\gamma(N+n)))e^{-1/\gamma}\,\mathbf{P}\{\zeta_S(\gamma) = N+n\}, \tag{13}$$

where $\zeta_S(\gamma) = \nu^{(1)}(\gamma) + \cdots + \nu^{(S)}(\gamma)$.

Let us prove that for sufficiently large N, n

$$\mathbf{P}\{\zeta_S(\gamma) = N+n\} \leqslant C_3 N^{-2}. \tag{14}$$

Denoting by $\varphi^{(r)}_{(\gamma)}(u)$ the characteristic function of the variable $\nu^{(1)}(\gamma)$ we represent the probability $\mathbf{P}\{\zeta_S(\gamma) = N+n\}$ by the invertion formula:

$$\mathbf{P}\{\zeta_S(\gamma) = N+n\}$$
$$= \frac{dB}{2\pi N^2}\int\limits_{-\pi N^2/(dB)}^{\pi N^2/(dB)} \exp\left\{-\frac{iu(N+n)}{N^2}\right\}\left(\varphi^{(r)}_{(\gamma)}\left(\frac{Bu}{N^2}\right)\right)^S du. \tag{15}$$

Considering the relation

$$\left|\varphi^{(r)}_{(\gamma)}(u)\right|^S = \left|\frac{R(1/(\gamma n)+iu)}{R(1/(\gamma n))}\right|^S$$

and using (10), (11) we obtain by analogy with (2.5.25) that for any fixed u

$$\left|\varphi^{(r)}_{(\gamma)}(u)\right|^S \leqslant C_4|\varphi^{(r)}(u)|^S. \tag{16}$$

In the proof of Lemma 1 we saw that if $|u| \leqslant \varepsilon N^2$ then the inequality (4) holds and if $\varepsilon N^2 < |u| \leqslant \pi N^2/(dB)$ then $|\varphi^{(r)}(u)| \leqslant e^{-C_5}$. Using these estimates and representing the integral which stands in the right-hand side of the equality (15) as the sum of two integrals we obtain from (16) that

$$\mathbf{P}\{\zeta_S(\gamma) = N+n\} \leqslant C_6 N^{-2}\int\limits_0^\infty \exp\{-C_7\sqrt{u}\}du + C_8N^2e^{-C_9N}.$$

This inequality implies (14). Using (8) and (12)–(14) we get

$$P_{S1}(n) \leqslant C_{10}N^{-2}e^{-1/\gamma} = o(S/n^{3/2}). \tag{17}$$

Now consider $P_{S2}(n)$. By (7)

$$P_{S2}(n) = \sum_{0\leqslant k\leqslant n(1-\gamma)} \mathbf{P}\left\{\nu^{(S)}_{(r)} = n-k\right\}$$
$$\times\,\mathbf{P}\left\{\nu^{(1)}_{(r)} + \cdots + \nu^{(S-1)}_{(r)} = N+k,\ \nu^{(i)}_{(r)} \leqslant \gamma n,\ i \leqslant S-1\right\}. \tag{18}$$

We note that

$$\mathbf{P}\left\{\nu^{(1)}_{(r)} + \cdots + \nu^{(S-1)}_{(r)} = N+k,\ \nu^{(i)}_{(r)} \leqslant \gamma n,\ i \leqslant S-1\right\}$$
$$= \mathbf{P}\left\{\nu^{(1)}_{(r)} + \cdots + \nu^{(S-1)}_{(r)} = N+k \mid \nu^{(i)}_{(r)} \leqslant \gamma n,\ i \leqslant S-1\right\}\mathbf{P}^{S-1}\left\{\nu^{(1)}_{(r)} \leqslant \gamma n\right\}. \tag{19}$$

Let $\nu_{(r)}^{(1)}(\gamma n), \dots, \nu_{(r)}^{(S)}(\gamma n)$ be auxiliary independent distributed variables such that

$$\mathsf{P}\left\{\nu_{(r)}^{(1)}(\gamma n) = k\right\} = \mathsf{P}\{\nu^{(1)} = k |\ \nu^{(1)} \leqslant \gamma n,\ \nu^{(1)} \neq r+1\}.$$

Also, let $\zeta_{S-1}^{(r)}(\gamma n) = \nu_{(r)}^{(1)}(\gamma n) + \dots + \nu_{(r)}^{(S-1)}(\gamma n)$. From (19) we see that

$$\mathsf{P}\left\{\nu_{(r)}^{(1)} + \dots + \nu_{(r)}^{(S-1)} = N+k,\ \nu_{(r)}^{(i)} \leqslant \gamma n,\ i \leqslant S-1\right\}$$
$$= \mathsf{P}\left\{\zeta_{S-1}^{(r)}(\gamma n) = N+k\right\} \mathsf{P}^{S-1}\left\{\nu_{(r)}^{(1)} \leqslant \gamma n\right\}.$$

This together with (11), (18) implies that as $N, n \to \infty$, $n/N^2 \to \infty$

$$P_{S2}(n) = (1+o(1)) \sum_{0 \leqslant k < n(1-\gamma)} \mathsf{P}\left\{\nu_{(r)}^{(S)} = n-k\right\}$$
$$\times \mathsf{P}\left\{\zeta_{S-1}^{(r)}(\gamma n) = N+k\right\}. \tag{20}$$

It is easy to see that

$$\mathsf{P}\left\{\zeta_{S-1}^{(r)}(\gamma n) > \gamma^{-1}N^2\right\} \leqslant (S-1)\,\mathsf{P}\left\{\nu_{(r)}^{(1)}(\gamma n) > \gamma^{-1}N^2\right\}$$
$$+ \mathsf{P}^{S-1}\left\{\nu_{(r)}^{(1)}(\gamma n) \leqslant \gamma^{-1}N^2\right\} \mathsf{P}\Big\{\zeta_{S-1}^{(r)}(\gamma n) > \gamma^{-1}N^2|$$
$$|\ \nu_{(r)}^{(1)}(\gamma n) \leqslant \gamma^{-1}N^2,\ i \leqslant S-1\Big\} \leqslant (S-1)\,\mathsf{P}\Big\{\nu_{(r)}^{(1)}(\gamma n)$$
$$> \gamma^{-1}N^2\Big\} + \mathsf{P}\left\{\zeta_{S-1}^{(r)}(\gamma^{-1}N^2) > \gamma^{-1}N^2\right\}. \tag{21}$$

Note, that

$$\mathsf{P}\left\{\nu_{(r)}^{(1)}(\gamma n) > \gamma^{-1}N^2\right\} \leqslant \sum_{k > \gamma^{-1}N^2} \mathsf{P}\left\{\nu_{(r)}^{(1)}(\gamma n) = k\right\}$$
$$\leqslant C_{11}\,\mathsf{P}\left\{\nu_{(r)}^{(1)} > \gamma^{-1}N^2\right\}.$$

From this and (11) we obtain

$$(S-1)\,\mathsf{P}\left\{\nu_{(r)}^{(1)}(\gamma n) > \gamma^{-1}N^2\right\} \leqslant C_{12}\gamma^{1/2}. \tag{22}$$

Using (10) for sufficiently large N, n we get

$$\mathsf{E}\nu_{(r)}^{(1)}(\gamma^{-1}N^2) \leqslant C_{13} \sum_{k \leqslant \gamma^{-1}N^2} k\,\mathsf{P}\left\{\nu_{(r)}^{(1)} = k\right\} \leqslant C_{14}\gamma^{-1/2}N. \tag{23}$$

This and the Chebyshev inequality imply that

$$\mathsf{P}\left\{\zeta_{S-1}^{(r)}(\gamma^{-1}N^2) > \gamma^{-1}N^2\right\} \leqslant C_{15}\gamma^{1/2}; \tag{24}$$

therefore, from (21) and (22) we obtain

$$\mathsf{P}\left\{\zeta_{S-1}^{(r)}(\gamma n) > \gamma^{-1}N^2\right\} \leqslant C_{16}\gamma^{1/2}. \tag{25}$$

The relation (20) can be represented as

$$P_{S2}(n) = S_1 + S_2 + S_3, \tag{26}$$

where

$$S_i = (1+o(1))\sum_{K_i} \mathbf{P}\left\{\nu_{(r)}^{(S)} = n-k\right\} \mathbf{P}\left\{\zeta_{S-1}^{(r)}(\gamma n) = N+k\right\}, \tag{27}$$

$$K_1 = \{k:\ 0 \leqslant k \leqslant \gamma^{-1}N^2 - N\};$$
$$K_2 = \{k:\ \gamma^{-1}N^2 - N < k \leqslant n(1-\gamma^{1/6})\};$$
$$K_3 = \{k:\ n(1-\gamma^{1/6}) < k \leqslant n(1-\gamma)\}.$$

If $k \leqslant \gamma^{-1}N^2 - N$ then from Lemma 1.3.12, (1.3) and (27) we get

$$S_1 = d((1-q_r(1))n\sqrt{2\pi Bn})^{-1}(1+o(1))\, \mathbf{P}\left\{\zeta_{S-1}^{(r)}(\gamma n) \leqslant \gamma^{-1}N^2\right\}.$$

From this and (25) we obtain

$$S_1 = d((1-q_r(1))n\sqrt{2\pi Bn})^{-1}(1+o(1)). \tag{28}$$

Using (1.3), (25), (27) and Lemma 1.3.12 we find for $k \in K_2$

$$S_2 \leqslant C_{17}(n\gamma^{1/6})^{-3/2}\, \mathbf{P}\left\{\zeta_{S-1}^{(r)}(\gamma n) > \gamma^{-1}N^2\right\} = o(n^{-3/2}). \tag{29}$$

If $k \in K_3$ then using again (1.3), (27) and Lemma 1.3.12 we have

$$S_3 \leqslant C_{18}(\gamma n)^{-3/2}\, \mathbf{P}\left\{\zeta_{S-1}^{(r)}(\gamma n) > n(1-\gamma^{1/6})\right\}. \tag{30}$$

Estimating $\mathbf{E}\,\nu_{(r)}^{(1)}(\gamma n)$ and using the Chebyshov inequality it is easy to find by analogy with (23) and (24) that

$$\mathbf{P}\left\{\zeta_{S-1}^{(r)}(\gamma n) > n(1-\gamma^{1/6})\right\} \leqslant C_{19}(\gamma/n)^{1/2}.$$

This and (30) imply the relation $S_3 = o(n^{-3/2})$ and from (26), (28), (29) we obtain

$$P_{S2}(n) = d(2\pi Bn^3)^{-1/2}(1+o(1)). \tag{31}$$

To estimate $P_{S3}(n)$ we note that

$$\begin{aligned} P_{S3}(n) \leqslant N(N-1)2^{-1} &\sum_{k\leqslant N+n(1-2\gamma)} \mathbf{P}\left\{\zeta_{S-2}^{(r)} = k\right\} \\ \times\, \mathbf{P}&\left\{\nu_{(r)}^{(S-1)} + \nu_{(r)}^{(S)} = N+n-k,\ \nu_{(r)}^{(S-1)} > \gamma n,\ \nu_{(r)}^{(S)} > \gamma n\right\}. \end{aligned} \tag{32}$$

It is clear that the last expression is less than

$$C_{20}N^2 \sum_{k\leqslant N+n(1-2\gamma)} \mathbf{P}\left\{\zeta_{S-2}^{(r)} = k\right\}$$
$$\times\left(\sum_K \mathbf{P}\left\{\nu_{(r)}^{(S-1)} = i\right\} \mathbf{P}\left\{\nu_{(r)}^{(S)} = N+n-k-i\right\}\right),$$

where $K = \{i:\ \gamma n < i < N-k+n(1-\gamma)\}$. Using Lemma 1.3.12 and (11) we obtain

$$\sum_K \mathbf{P}\left\{\nu_{(r)}^{(S-1)} = i\right\} \mathbf{P}\left\{\nu_{(r)}^{(S)} = N+n-k-i\right\}$$
$$\leqslant C_{21}(\gamma n)^{-3/2}\, \mathbf{P}\left\{\nu_{(r)}^{(S)} > \gamma n\right\} \leqslant C_{22}(\gamma n)^{-2}.$$

From this and (32) it follows that

$$P_{S3}(n) \leqslant C_{23}N^2(\gamma n)^{-2} = o(S/n^{3/2});$$

therefore, from (7), (17) and (31) we obtain the assertion of Lemma 2.

4. Proofs of the main results

In this Section we will prove Theorems 1.1–1.4.

It is well known that if $q_r(\lambda) \to 0$ then for non-negative integers k

$$\binom{N}{k} q_r^k(\lambda)(1-q_r(\lambda))^{N-k} = \frac{(Nq_r(\lambda))^k}{k!}e^{-Nq_r(\lambda)}(1+o(1)) \tag{1}$$

uniformly in $(k-Nq_r(\lambda))/\sqrt{Nq_r(\lambda)}$ lying in any finite interval. Under the hypothesis of Theorem 1.1 the inequality $r > j$ holds and (1.5) implies that $q_r(\lambda) \to 0$. Therefore the assertion of Theorem 1.1 follows from Lemmas 1.1, 2.3, 2.3.2 and the equality (1).

According to the normal approximation of the binomial distribution under the condition $Nq_r(\lambda)(1-q_r(\lambda)) \to \infty$

$$\begin{aligned}&\binom{N}{k} q_r^k(\lambda)(1-q_r(\lambda))^{N-k}\\ &= \frac{1+o(1)}{\sqrt{2\pi Nq_r(\lambda)(1-q_r(\lambda))}}\exp\left\{-\frac{(k-Nq_r(\lambda))^2}{2Nq_r(\lambda)(1-q_r(\lambda))}\right\}\end{aligned} \tag{2}$$

uniformly in $(k-Nq_r(\lambda))/\sqrt{Nq_r(\lambda)(1-q_r(\lambda))}$ lying in any finite interval. If $N, n \to \infty$ in such a way that $n/N \geqslant C > 0$, $n/N^2 \to 0$, then the assertion of Theorem 1.2 can be obtained from Lemmas 1.1, 2.2, 2.3.2 and the relation (2). Let $0 < C_1 \leqslant n/N^2 \leqslant C_2 < \infty$. Putting $\lambda = 1$ we deduce from Lemmas 1.1, 3.1, 2.3.3 and (2) that

$$\mathsf{P}\{\mu_r(\mathfrak{F}) = k\} = (2\pi Nq_r(1)(1-q_r(1)))^{-1/2}e^{-u_r^2/2}(1+o(1)) \tag{3}$$

uniformly in $u_r = (k-Nq_r(1))/\sqrt{Nq_r(1)(1-q_r(1))}$ lying in any finite interval. Using the relations (1.4)–(1.7) and (2.6) it is not hard to see that if $n/N^2 \geqslant C_1 > 0$ then the relation (3) is equivalent to the assertion of Theorem 1.2. To conclude the proof it remains to note that under the condition $n/N^2 \to \infty$ the equality (3) is the corollary of (2) and Lemmas 1.1, 3.2, 2.3.4.

If the hypotheses of Theorem 1.3 are satisfied, its assertion can be easily obtained from (2) and Lemmas 1.1, 2.3, 2.3.2. We note that the condition $N\lambda^r \to \infty$ is necessary to use the relation (2).

If $n/N \to 0$ then the assertion of Theorem 1.4 follows from (1) and Lemmas 1.1, 2.3 and 2.3.2. If $n/N \geqslant C > 0$, $n/N^2 \to 0$ then Theorem 1.4 follows from (1) and Lemmas 1.1, 2.2 and 2.3.2. In the case $0 < C_1 \leqslant n/N^2 \leqslant C_2 < \infty$ we can use (1) and Lemmas 1.1, 3.1, 2.3.3. Let, finally $n/N^2 \to \infty$. Then to prove Theorem 1.4 we can use (1) and Lemmas 1.1, 3.2, 2.3.4.

5. Additions and references

Chapter 3 relies mainly on the paper [67]. Limit distributions of the number of trees of a given size in a random forest were first obtained in [52], where forests from $\mathfrak{F}'_{N,n}$ (Example 1.1.1) with the uniform probability distribution were considered. Proofs in [52] are based on the use of the generalized allocation scheme without relying on the connection of forests with random processes. For example, for the random variable $\mu_r(\mathfrak{F}') = \mu_r(\mathfrak{F}'_{N,n})$ the following results were obtained.

Theorem 1. *Let $N, n \to \infty$ so that $\alpha = 3n^2/(2N) \leqslant C$, where C is some positive constant. Then for any fixed $k = 0, 1, \dots$*

$$\mathbf{P}\{\mu_0(\mathfrak{F}') - N + n = k\} = \frac{\alpha^k}{k!}e^{-\alpha} + o(1),$$

$$\mathbf{P}\left\{\frac{n - \mu_1(\mathfrak{F}')}{2} = k\right\} = \frac{\alpha^k}{k!}e^{-\alpha} + o(1).$$

Let

$$\sigma_{rr}^2(x) = p'_r(x)(1 - p'_r(x) - (n/N - r)^2 p'_r(x)N^3/(n(n+N)^2)),$$

where

$$p'_r(x) = \frac{(r+1)^r}{(r+1)!}x^r e^{-\theta(x)}, \quad r = 0, 1, \dots, \tag{1}$$

$$\theta(x) = \sum_{k=1}^{\infty} k^{k-1}x^k/k!, \quad x = \frac{n}{N+n}\exp\left\{-\frac{n}{N+n}\right\}.$$

Theorem 2. *Let $N, n \to \infty$ so that $n^2/N \to \infty$. Then*

$$\mathbf{P}\left\{\frac{\mu_1(\mathfrak{F}') - Np'_1(x)}{\sigma_{11}(x)\sqrt{N}} \leqslant z\right\} \to \frac{1}{\sqrt{2\pi}}\int_{-\infty}^{z} e^{-y^2/2}dy.$$

Theorem 3. *Let $N, n \to \infty$ so that $n/N \to 0$, $n^3/N^2 \to \infty$. If $\alpha = (n - Np'_1(x))/2$ then for non-negative integers k*

$$\mathbf{P}\{\mu_1(\mathfrak{F}') = n - 2k\} = \frac{1}{2}\frac{\alpha^k e^{-\alpha}}{k!}(1 + o(1)),$$

$$\mathbf{P}\{\mu_1(\mathfrak{F}') = n - 2k - 1\} = \frac{1}{2}\frac{\alpha^k e^{-\alpha}}{k!}(1 + o(1))$$

uniformly in non-negative integers k such that $(k - \alpha)/\sqrt{\alpha}$ lies in any finite interval.

To prove Theorems 1–3 the paper [52] made wide use of the explicit form of the distribution (1). Therefore the results do not directly follow from Theorems 1.1–1.4. Thus, when the specific distribution of the number of descendants of one particle (1.1) in the process G is known, one can get specifications of the limit Theorems 1.1–1.4.

It follows from Theorems 1.3 and 1.4 that local limit theorems were obtained for the random variable $\mu_r(\mathfrak{F}')$ at all r values and different variants of N and n tendency to infinity. The value $\mu_1(\mathfrak{F}')$ for which only the integral Theorem 2 is valid in the case $n^2/N \to \infty$, $n^3/N^2 \leqslant C < \infty$ is the exception. A local variant of the integral theorem was considered in [56], where the following results were obtained.

Theorem 4. *Let $N, n \to \infty$ so that $n^2/N \to \infty$, $n^3/N^2 \to 0$. Then for $\alpha = 3n^2/(2N)$*

$$\mathsf{P}\left\{\mu_1(\mathfrak{F}') = n - 2k\right\} = \frac{\alpha^k}{k!} e^{-\alpha}(1 + o(1))$$

uniformly in $(k-\alpha)/\sqrt{\alpha}$ lying in any finite interval.

Theorem 5. *Let $N, n \to \infty$ so that $n^3/N^2 \to \gamma$, where γ is some positive constant. Then for $\alpha = 3n^2/(2N)$*

$$\mathsf{P}\left\{\mu_1(\mathfrak{F}') = n - 2k\right\} = \frac{1}{2}\frac{\alpha^k e^{-\alpha}}{k!}\left(1 + e^{-16\gamma/3}\right)(1 + o(1)),$$
$$\mathsf{P}\left\{\mu_1(\mathfrak{F}') = n - 2k - 1\right\} = \frac{1}{2}\frac{\alpha^k e^{-\alpha}}{k!}\left(1 - e^{-16\gamma/3}\right)(1 + o(1))$$

uniformly in non-negative integers k such that $(k-\alpha)/\sqrt{\alpha}$ lies in any finite interval.

An interesting fact in the limit behaviour of the variable $\mu_1(\mathfrak{F})$ is that as $n^2/N \to \infty$, $n^3/N^2 \leqslant C < \infty$ there occurs integral convergence of the $\mu_1(\mathfrak{F}')$ distribution to the normal law, but the limit distribution in the local sense is a law different from the normal law. Similar behaviour is displayed by the random variable equal to the number of cells with exactly one particle under equiprobable allocation of n different particles in N cells [43].

It follows from Theorem 5 that if $N, n \to \infty$ and the variable n^3/N^2 is limited, then

$$\sum_{k=0}^{\infty} \mathsf{P}\left\{\mu_1(\mathfrak{F}') = n - 2k\right\} = 2^{-1}(1 + \exp\{-16n^3/(3N^2)\}) + o(1),$$
$$\sum_{k=0}^{\infty} \mathsf{P}\left\{\mu_1(\mathfrak{F}') = n - 2k - 1\right\} = 2^{-1}(1 - \exp\{-16n^3/(3N^2)\}) + o(1). \tag{2}$$

It follows from Theorems 1, 4 that if $n^3/N^2 \to 0$, then the distribution of the variable $n - \mu_1(\mathfrak{F}')$ asymptotically lies on the lattice of even non-negative numbers. As becomes clear from (2) however, if n^3/N^2 ratio is limited, then the distribution of $n - \mu_1(\mathfrak{F}')$ moves to a different lattice. This makes the integral Theorem 2 valid for a wider parameter variation domain as compared with the local Theorem 3.

Article [70] gives estimates of the rate of convergence of $\mu_r(\mathfrak{F}')$ distributions to the limit distributions.

Theorem 6. *Let $N, n \to \infty$ so that $n/N \to 0$, $Np_r'(x) \to \infty$. Let $u_r = (k - Np_r'(x))/(\sigma_{rr}\sqrt{N})$, $u = (k - Np_r'(x))/\sqrt{Np_r'(x)}$, where k, r are integer positive numbers, $r \geqslant 2$. If $u_r^6 = o(Np_r'(x))$, $Np_r'(x)u_r^2/n \leqslant C < \infty$, $\beta = r^2 Np_r'(x)/n$,*

$u^2\beta \to 0$, *then*

$$\mathbf{P}\{\mu_r(\mathfrak{F}') = k\} = (\sigma_{rr}(x)\sqrt{2\pi N})^{-1}$$
$$\times\left(1 + \frac{3u_r - u_r^2}{6\sqrt{Np'_r(x)}} + o\left(\frac{1+u_r^6}{\sqrt{Np'_r(x)}}\right)\right),$$
$$\mathbf{P}\{\mu_r(\mathfrak{F}') = k\} = (k!)^{-1}(Np'_r(x))^k \exp\{-Np'_r(x)\}$$
$$\times(1 + \beta 2^{-1}(1-u^2) + O(p'_r(x)(1+u^2) + \beta^2 u^4 + (1+|u|\sqrt{Np'_r(x)})n^{-1})).$$

Theorem 6 in [70] was emloyed to estimate the distance in variation between $\mu_r(\mathfrak{F}')$ distributions and the limit distributions.

These results can be used to identify the limit distribution closest to $\mu_r(\mathfrak{F}')$ in the sense of this distance with respect to the behaviour of N, n. The theorems proved in [70] are similar to the results about the convergence to the limit laws of the number of cells with exactly r particles under equiprobable allocation of n different particles in N different cells [43].

Denote

$$\Pi(k,\alpha) = \alpha^k/(k!e^\alpha), \quad k = 0,1,\ldots;$$
$$N(y,a,\sigma^2) = (\sigma\sqrt{2\pi})^{-1}\exp\{-(y-a)^2/(2\sigma^2)\}.$$

Introduce the distance between $\mu_r(\mathfrak{F}')$ distributions and the limit distributions

$$\rho_{1r} = \sum_{k=0}^{\infty} \left|\mathbf{P}\{\mu_r(\mathfrak{F}') = k\} - \Pi(k, Np'_r(x))\right|,$$

$$\rho_{2r} = \sum_{k=0}^{\infty} \left|\mathbf{P}\{\mu_r(\mathfrak{F}') = k\} - N(k, Np'_r(x), N\sigma_{rr}^2(x))\right|.$$

Theorem 7. *Let $N, n \to \infty$ so that $n/N \to 0$, $Np'_r(x) \to \infty$. Then for any fixed $r \geqslant 2$*

$$\begin{aligned} \rho_{1r} &= b_1\beta(1+o(1)) \\ \rho_{2r} &= b_2(Np'_r(x))^{-1/2}(1+o(1)), \end{aligned} \tag{3}$$

where $\beta = r^2 Np'_r(x)/n$, $b_1 = \sqrt{2/(\pi e)}$, $b_2 = (1+4e^{-3/2})/(3\sqrt{2\pi})$.

It is easy to see from the relations (1) and (3) that as n/N increases, ρ_{1r} monotonically increases, and ρ_{2r} monotonically decreases. Setting ρ_{1r} equal to ρ_{2r} we find the point γ where normal and Poisson approximations have equal precision from the point of view of the distance in variation:

$$\gamma = \frac{1}{r+1}\left(\frac{r!}{r^{4/3}}\right)^{1/(r-1)}\left(\frac{1}{n}\left(\frac{b_2}{b_1}\right)^2\right)^{1/(3(r-1))}(1+o(1)).$$

Of the two limit distributions, a Poisson distribution yields a closer approximation at $n/N < \gamma$, and a normal distribution at $n/N > \gamma$.

Results describing the limit behaviour of the number of trees of a given size in a random forest of non-rooted trees with labelled vertices were obtained in [9,10].

CHAPTER 4

LIMIT DISTRIBUTIONS OF THE HEIGHT OF A RANDOM FOREST

1. Problem statement and summary of results

It is easy to see that the equation (1.3.55) is the corollary of Theorem 1.4.1. Therefore to investigate the limit behaviour of the height τ of a random forest from $\mathfrak{F}_{N,n}$ we can use Lemmas 1.3.10 and 1.3.11. We remind the reader of some denotations. The set $\mathfrak{F}_{N,n}$ is connected with the branching process G beginning with N particles. These particles are the initial particles of the processes $G^{(1)},\dots,G^{(N)}$, where $\bigcup_{i=1}^{N} G^{(i)} = G$. Let $\nu^{(i)}$ be the total progeny of the processes $G^{(i)}$, $i=1,\dots,N$ and $\nu_N = \nu^{(1)}+\dots+\nu^{(N)}$. We denote by $\mu^{(i)}(t)$ the numbers of particles of the t-th generation of the process $G^{(i)}$ and $\mu(t)=\mu^{(1)}(t)+\dots+\mu^{(N)}(t)$. Let $\nu^{(1)}(t),\dots,\nu^{(N)}(t)$ be independent identically distributed random variables such that

$$\mathsf{P}\{\nu^{(i)}(t)=k\} = \mathsf{P}\{\nu^{(i)}=k|\mu^{(i)}(t)=0\}, \tag{1}$$

where $i=1,\dots,N$, $k=1,2,\dots$ and let $\zeta_N^{(t)} = \nu^{(1)}(t)+\dots+\nu^{(N)}(t)$. By Lemmas 1.3.10 and 1.3.11

$$\mathsf{P}\{\tau<t\} = \mathsf{P}\{\mu(t)=0\}\,\mathsf{P}\{\zeta_N^{(t)}=N+n\}/\,\mathsf{P}\{\nu_N=N+n\}, \tag{2}$$

$$\mathsf{P}\{\tau<t\} = 1-\frac{\mathsf{P}\{\mu(t)>0\}\,\mathsf{P}\{\nu_N=N+n|\mu(t)>0\}}{\mathsf{P}\{\nu_N=N+n\}}. \tag{3}$$

These relations will be used to prove Theorems 1–5 on the limit behaviour of the random forest height. We will consider the branching process G with the offspring distribution

$$p_k(\lambda) = \lambda^k p_k/F(\lambda), \quad k=0,1,\dots, \tag{4}$$

$0<\lambda\leqslant 1$ and

$$F(z)=\sum_{k=0}^{\infty} p_k z^k, \tag{5}$$

$F(1)=1$, $F'(1)=1$, $F''(1)=B$, $p_0>0$, the maximal span of the distribution (4) is d. We denote by j the least positive integer such that $p_j>0$ and let l be the least natural number not divided by j for which $p_{j+l}>0$; if such l does not exist, we put $l=0$. Let $\lambda=\lambda(N,n)$ be determined by the relation

$$\frac{\lambda F'(\lambda)}{F(\lambda)} = \frac{n}{N+n}, \tag{6}$$

and let m and B_λ be the mathematical expectation and variance of the distribution (4) respectively.

Theorem 1. *Let $N, n \to \infty$ in such a way that $n/N \to 0$, $n^2/N \to \infty$, $N\lambda^{j+l} \to \infty$ and let $t = t(N,n)$ be chosen such that $Nm^t \to \infty$, Nm^{t+1} are bounded. Then*

$$\mathsf{P}\{\tau = t\} = \exp\left\{-Nm^{t+1}\right\} + o(1),$$
$$\mathsf{P}\{\tau = t+1\} = 1 - \exp\left\{-Nm^{t+1}\right\} + o(1).$$

Remark 1. If there exists a positive constant C such that $Nm^{t+1} > C$, then $Nm^t \to \infty$. From the hypotheses of Theorem 1 it follows that t can be a constant. We will obtain some of the sufficient conditions for $t = t(N,n)$ in Theorem 1 in Section 4.

We denote by $F_\lambda(z)$ the generating function of the distribution (4). For positive integers r, let the expression $F_\lambda^{*r}(z)$ mean the r-th iteration of the function $F_\lambda(z)$; hence

$$F_\lambda^{*1}(z) = F_\lambda(z), \quad F_\lambda^{*2}(z) = F_\lambda(F_\lambda(z)), \quad F_\lambda^{*3}(z) = F_\lambda(F_\lambda(F_\lambda(z))), \ldots$$

Theorem 2. *Let $N, n \to \infty$ in such a way that n takes values which are divided by d, $n/N \to b$ and let $t = t(N,n)$ be chosen such that $Nm^t \to \beta$, where b, β are positive constants. Then for any fixed $k = 0, \pm 1, \pm 2, \ldots$*

$$\mathsf{P}\{\tau < t+k\} \to \exp\left\{-K\beta(b/(1+b))^k\right\},$$

where

$$K = \lim_{r\to\infty} \frac{1 - F_\lambda^{*r}(0)}{m^r}. \tag{7}$$

Theorem 3. *Let $N, n \to \infty$ in such a way that n takes values which are divided by d, $n/N \to \infty$, $n/N^2 \to 0$, $F'''(1) < \infty$. Then for any fixed x*

$$\mathsf{P}\left\{\tau \ln(1 + N/n) - \ln(2N^2/(Bn)) \leqslant x\right\} \to e^{-e^{-x}}.$$

Theorem 4. *Let $N, n \to \infty$ in such a way that n takes values which are divided by d, $N/\sqrt{Bn} \to z$. Then for any fixed $x > 0$ uniformly with respect to z in any interval of the form $0 < z_0 \leqslant z \leqslant z_1 < \infty$*

$$\mathsf{P}\left\{(B/n)^{1/2}\tau \leqslant x\right\} \to e^{z^2/2}\left(z\sqrt{2\pi}\right)^{-1} \int_{-\infty}^{\infty} \exp\left\{-iu - zf(x,u)/x\right\} du,$$

where

$$f(x,u) = \frac{x\sqrt{-2iu}(1 + \exp\{-x\sqrt{-2iu}\})}{1 - \exp\{-x\sqrt{-2iu}\}}. \tag{8}$$

Theorem 5. *Let $n \to \infty$ in such a way that n takes values which are divided by d, $n/N^2 \to \infty$. Then for any fixed $x > 0$*

$$\mathsf{P}\left\{(B/n)^{1/2}\tau < x\right\} \to \sum_{k=-\infty}^{\infty} (1 - k^2x^2)\exp\left\{-k^2x^2/2\right\}.$$

2. The limit behaviour of auxiliary probabilities in the subcritical case

To prove Theorems 1.1–1.3 in Section 4 we will use the relation (1.2) under the condition $\lambda < 1$. Therefore the process G is subcritical. The behaviour of the probability $\mathbf{P}\{\nu_N = N+n\}$ was considered in Section 2.3. This and (1.2) mean that to prove the results in the subcritical case it suffices to have the asymptotics of $\mathbf{P}\{\mu(t)=0\}$ and $\mathbf{P}\{\zeta_N^{(t)} = N+n\}$.

Lemma 1. *If $N \to \infty$ then the next assertions are valid:*

1. *If $0 < m \leqslant C < 1$ and $t = t(\lambda)$ be chosen such that $Nm^t \to \beta$, where β is some positive constant then*

$$\mathbf{P}\{\mu(t)=0\} = e^{-K\beta} + o(1),$$

where the constant K is determined by the equality (1.7) *and $K=1$ as $m \to 0$.*

2. *If $m \to 1$, $N(1-m) \to \infty$, $F'''(1) < \infty$,*

$$t = (-x + \ln(B_\lambda/(2N(1-m))))/\ln m + O(1), \tag{1}$$

where x is a positive constant, then

$$\mathbf{P}\{\mu(t)=0\} = e^{-e^{-x}} + o(1).$$

Proof. Let $m \leqslant C < 1$. Then the processes $G^{(1)}, \dots, G^{(N)}$ are subcritical. Since $F''_\lambda(1) < \infty$, it follows that for $G^{(1)}$ the hypotheses of Theorem 1.3.2 are true; therefore

$$\mathbf{P}\{\mu^{(1)}(t) > 0\} = Km^t(1+o(1)), \tag{2}$$

where K is given by (1.7). Note that

$$\mathbf{P}\{\mu(t)=0\} = \left(\mathbf{P}\{\mu^{(1)}=0\}\right)^N = \left(1 - \mathbf{P}\{\mu^{(1)} > 0\}\right)^N; \tag{3}$$

hence (2), (3) and Corollary 1.3.1 imply the first assertion of Lemma 1. If $m \to 1$, then to prove the second assertion of the Lemma it suffices to use (3) and Theorem 1.3.4.

Further we will need to have the auxiliary assertion (Lemma 2) which connects some characteristics of the process G with similar characteristics of the process G^* with the generating function (1.5). Thus the process G^* consists of N critical processes. Let $\mu_r(t)$, $\mu_r^*(t)$ be the number of particles at the instant t having exactly r direct descendants in the processes G and G^* respectively and let ν_N^* be the total progeny of the process G^*. We denote by M a matrix of the size $(n+1)\times(n+1)$ composed of non-negative integers.

Lemma 2. *For any matrix M and any n such that $\mathbf{P}\{\nu_N^* = N+n\} > 0$*

$$\mathbf{P}\{||\mu_r(t)|| = M \mid \nu_N = N+n\} = \mathbf{P}\{||\mu_r^*(t)|| = M \mid \nu_N^* = N+n\}.$$

Proof. Let $M = ||m_r(t)||$ and $n_t = m_0(t) + \dots + m_n(t)$. For the equality $||\mu_r^*(t)|| = M$ to have positive probability it is necessary that

$$n_0 = N, \quad n_t = m_1(t-1) + 2m_2(t-1) + \dots + nm_n(t-1),$$
$$t = 1, \dots, n; \quad m_1(n) + 2m_2(n) + \dots + nm_n(n) = 0,$$
$$n_0 + n_1 + \dots + n_n = N+n.$$

We say that the matrix M is admissible if these conditions hold. For an admissible matrix

$$\mathbf{P}\{||\mu_r^*(t)|| = M\} = \prod_{t=0}^{n} \frac{n! p_0^{m_0(t)} \dots p_n^{m_n(t)}}{m_0(t)! \dots m_n(t)!}.$$

Therefore by (1.4)

$$\mathbf{P}\{||\mu_r(t)|| = M\} = \frac{\lambda^n}{F^{N+n}(\lambda)} \mathbf{P}\{||\mu_r^*(t)|| = M\}. \tag{4}$$

From Lemma 1.3.4 it follows by analogy with (2.2.4) that

$$\mathbf{P}\{\nu_N = N + n\} = \frac{\lambda^n}{F^{N+n}(\lambda)} \mathbf{P}\{\nu_N^* = N + n\}.$$

This together with (4) implies the assertion of Lemma 2 for admissible matrices. Lemma 2 is also true for inadmissible matrices, because in this case both probabilities in the assertion are equal to zero.

Now we consider the limit behaviour of the sum $\zeta_N^{(t)} = \nu^{(1)}(t) + \dots + \nu^{(N)}(t)$. Let $\varphi_\zeta^{(t)}(u)$ be the characteristic function of the random variable $(\zeta_N^{(t)} - aN)/(\sigma\sqrt{N})$, where

$$a = \mathbf{E}\,\nu^{(1)} = 1/(1-m), \quad \sigma^2 = \mathbf{D}\,\nu^{(1)} = B_\lambda/(1-m)^3. \tag{5}$$

Lemma 3. *Let $N, n \to \infty$ in such a way that $n/N^2 \to 0$, $N\lambda^j \to \infty$, $F'''(1) < \infty$ if $n/N \to \infty$ and let $t = t(N,n)$ be chosen such that $0 < C_1 \leqslant N\,\mathbf{P}\{\mu^{(1)}(t) > 0\} \leqslant C_2 < \infty$. Then*

$$\varphi_\zeta^{(t)}(u) \to e^{-u^2/2}$$

uniformly in u in any finite interval.

Proof. Let $\Psi^{(t)}(u)$ be the characteristic function of the random variable $\nu^{(1)}(t)$. Let also

$$f_N(w) = \sum_{k=0}^{\infty} \mathbf{P}\{\mu(0) = N,\ \nu_N = N + k\} w^{N+k},$$

$$f_{N,t}(w) = \sum_{k=0}^{\infty} \mathbf{P}\{\mu(0) = N,\ \mu(t) = 0,\ \nu_N = N + k\} w^{N+k}, \tag{6}$$

$$\Delta_{N,t}(w) = f_N(w) - f_{N,t}(w).$$

From this and (1.1) we obtain

$$\Psi^{(t)}(u) = \frac{f_{1,t}\left(e^{iu}\right)}{\mathbf{P}\{\mu^{(1)}(t) = 0\}} = \frac{f_1\left(e^{iu}\right) - \Delta_{1,t}\left(e^{iu}\right)}{\mathbf{P}\{\mu^{(1)}(t) = 0\}};$$

therefore

$$\varphi_\zeta^{(t)}(u) = \left[\exp\left\{-\frac{iau}{\sigma\sqrt{N}}\right\} \frac{f_1\left(e^{iu/(\sigma\sqrt{N})}\right) - \Delta_{1,t}\left(e^{iu/(\sigma\sqrt{N})}\right)}{\mathbf{P}\{\mu^{(1)}(t) = 0\}}\right]^N. \tag{7}$$

Using (2.3.7), (2.3.8), (2.3.3) and relation

$$f_1\left(e^{iu}\right) = \varphi(u) = \mathbf{E}\exp\{iu\nu^{(1)}\} \tag{8}$$

we can obtain by analogy with the proof of Lemma 2.3.1 that

$$\exp\left\{-\frac{iua}{\sigma\sqrt{N}}\right\} f_1\left(\exp\left\{\frac{iu}{\sigma\sqrt{N}}\right\}\right) = 1 - \frac{u^2}{2N} + o\left(\frac{1}{N}\right). \tag{9}$$

Let us consider the quantity $\Delta_{1,t}(\exp\{iu/(\sigma\sqrt{N})\})$. We will prove that

$$\Delta_{1,t}(\exp\{iu/(\sigma\sqrt{N})\}) = \mathbf{P}\{\mu^{(1)}(t) > 0\}(1 + o(1)). \tag{10}$$

We put $N = 1$, $w = \exp\{iu/(\sigma\sqrt{N})\}$ in the last of relations (6) and represent the resulting expression as the sum

$$\Delta_{1,t}(\exp\{iu/(\sigma\sqrt{N})\}) = S_1 + S_2, \tag{11}$$

where

$$\begin{aligned} S_1 &= \sum_{0\leqslant k\leqslant \varepsilon\sigma\sqrt{N}} \mathbf{P}\{\mu^{(1)}(t) > 0,\ \nu^{(1)} = k+1\}\exp\{iu(k+1)/(\sigma\sqrt{N})\}, \\ S_2 &= \sum_{k>\varepsilon\sigma\sqrt{N}} \mathbf{P}\{\mu^{(1)}(t) > 0,\ \nu^{(1)} = k+1\}\exp\{iu(k+1)/(\sigma\sqrt{N})\}; \end{aligned} \tag{12}$$

the positive constant ε will be chosen later. It is not hard to see that

$$S_1 = (1+\varepsilon_1) \sum_{0\leqslant k\leqslant \varepsilon\sigma\sqrt{N}} \mathbf{P}\{\mu^{(1)}(t) > 0,\ \nu^{(1)} = k+1\},$$

where ε_1 can be made arbitrarily small by the choice of sufficiently large N and sufficiently small ε. From this we obtain

$$\begin{aligned} S_1 = (1+\varepsilon_1)\,\mathbf{P}\{\mu^{(1)}(t) > 0\}\Big(1 - \big(\mathbf{P}\{\mu^{(1)}(t) > 0\}\big)^{-1} \\ \times \sum_{k>\varepsilon\sigma\sqrt{N}} \mathbf{P}\{\mu^{(1)}(t) > 0,\ \nu^{(1)} = k+1\}\Big). \end{aligned} \tag{13}$$

By Lemma 2

$$\begin{aligned} \sum_{k>\varepsilon\sigma\sqrt{N}} \mathbf{P}\{\mu^{(1)}(t) > 0,\ \nu^{(1)} = k+1\} = \mathbf{P}\left\{\mu_*^{(1)}(t) > 0\right\} \\ \times \sum_{k>\varepsilon\sigma\sqrt{N}} \mathbf{P}\left\{\nu_*^{(1)} = k+1 \mid \mu_*^{(1)}(t) > 0\right\} \mathbf{P}\{\nu^{(1)} = k+1\}/\,\mathbf{P}\left\{\nu_*^{(1)} = k+1\right\}, \end{aligned} \tag{14}$$

where $\mu_*^{(1)}(t)$, $\nu_*^{(1)}$ are the number of particles at the instant t and the total progeny of the critical branching process with one initial particle respectively. By (2.2.4)

$$\mathbf{P}\{\nu^{(1)} = k+1\}/\,\mathbf{P}\left\{\nu_*^{(1)} = k+1\right\} = \lambda^k/F^{k+1}(\lambda) \tag{15}$$

and by Theorem 1.3.3

$$\mathbf{P}\left\{\mu_*^{(1)}(t) > 0\right\} \leqslant C_3 t^{-1},$$

where the symbols $C_3, C_4, \ldots$ here and below denote positive constants. From this and (14), (15) we obtain

$$\sum_{k>\varepsilon\sigma\sqrt{N}} \mathbf{P}\{\mu^{(1)}(t)>0,\ \nu^{(1)}=k+1\} \leqslant C_4 t^{-1} \sum_{k>\varepsilon\sigma\sqrt{N}} (\lambda/F(\lambda))^k. \tag{16}$$

Now we need to get that for any fixed ε

$$\left(\mathbf{P}\{\mu^{(1)}(t)>0\}\right)^{-1} \sum_{k>\varepsilon\sigma\sqrt{N}} \mathbf{P}\{\mu^{(1)}(t)>0,\ \nu^{(1)}=k+1\} = o(1). \tag{17}$$

In the proof of Lemma 2.2.3 we saw that $\lambda/F(\lambda) < 1$; therefore the condition $N\,\mathbf{P}\{\mu^{(1)}(t)>0\} \geqslant C_1$ and (16) imply that

$$\begin{aligned}\left(\mathbf{P}\{\mu^{(1)}(t)>0\}\right)^{-1} &\sum_{k>\varepsilon\sigma\sqrt{N}} \mathbf{P}\{\mu^{(1)}(t)>0,\ \nu^{(1)}=k+1\} \\ &\leqslant C_5 N t^{-1} (\lambda/F(\lambda))^{\varepsilon\sigma\sqrt{N}} (1-\lambda/F(\lambda))^{-1}.\end{aligned} \tag{18}$$

If $0 < C_6 \leqslant \lambda \leqslant C_7 < 1$ then from (5), (18) and (1.6) we obtain that for some positive $q<1$

$$N t^{-1} (\lambda/F(\lambda))^{\varepsilon\sigma\sqrt{N}} < N q^{C_8\sqrt{N}} \to 0;$$

therefore (18) implies (17). Let $\lambda \to 0$. Using again (5), (18), (1.6), (2.3.4) and the relation $N\lambda^j \to \infty$ we get

$$N t^{-1} (\lambda/F(\lambda))^{\varepsilon\sigma\sqrt{N}} < N (C_9\lambda)^{C_{10}\sqrt{N\lambda^j}} < C_{11} N \lambda^j q^{C_{12}\sqrt{N\lambda^j}} \to 0$$

and (17) is valid. Finally, let $\lambda \to 1$. Using (2.2.15)–(2.2.18) we can find that

$$\begin{aligned}\sum_{k>\varepsilon\sigma\sqrt{N}} \mathbf{P}\{\mu^{(1)}(t)>0,\ \nu^{(1)}=k+1\} &\leqslant P_{\varepsilon\sigma\sqrt{N}} \\ &\leqslant C_{13}\frac{N}{n}\left(\frac{N}{\sqrt{n}}\right)^{-3/2} e^{-C_{14}N/\sqrt{n}}.\end{aligned} \tag{19}$$

Taking into consideration the condition $0 < C_1 \leqslant N\,\mathbf{P}\{\mu^{(1)}(t)>0\} \leqslant C_2 < \infty$ we see that (18) and (19) imply (17). From (13) and (17) we have

$$S_1 = \mathbf{P}\{\mu^{(1)}(t)>0\}(1+o(1)). \tag{20}$$

For these reasons and (18)

$$S_2 = o\left(\mathbf{P}\{\mu^{(1)}(t)>0\}\right)$$

and from (11), (20) we obtain (10). We can easily prove the assertion of Lemma 3 from (7), (9), (10) and the condition $0 < C_1 \leqslant N\,\mathbf{P}\{\mu^{(1)}(t)>0\} \leqslant C_2 < \infty$.

Remark 1. Let the conditions of Lemma 3 hold and $\lambda \to 0$. Then for any fixed u

$$\varphi_\zeta^{(t+1)}(u) \to e^{-u^2/2}, \quad \varphi_\zeta^{(t-1)}(u) \to e^{-u^2/2}.$$

Indeed, putting $t+1$ or $t-1$ instead of t in the proof of Lemma 3 and substituting the first multiplier of the right-hand side of the relation (18) for N/m or Nm we can easily obtain by Theorem 1.3.2 and Corollary 1.3.1 the assertion of Remark 1.

Lemma 4. *Let $N, n \to \infty$ in such a way that $n/N^2 \to 0$, $N\lambda^{j+l} \to \infty$, $F'''(1) < \infty$ if $n/N \to \infty$ and let $t = t(N,n)$ be chosen such that $0 < C_1 \leqslant N\,\mathbf{P}\{\mu^{(1)}(t) > 0\} \leqslant C_2 < \infty$. Then for non-negative h divisible by d*

$$\mathbf{P}\left\{\zeta_N^{(t)} = N + h\right\} = \frac{d(1+o(1))}{\sigma\sqrt{2\pi N}} \exp\left\{-\frac{(h-n)^2}{2\sigma^2 N}\right\}$$

uniformly in $(h-n)/(\sigma\sqrt{N})$ lying in any finite interval.

Proof. Using the inversion formula we represent the probability $\mathbf{P}\{\zeta_N^{(t)} = N + h\}$ as the integral

$$\mathbf{P}\left\{\zeta_N^{(t)} = N + h\right\} = \frac{d}{2\pi\sigma\sqrt{N}} \int\limits_{-\pi\sigma\sqrt{N}/d}^{\pi\sigma\sqrt{N}/d} e^{-izu}\varphi_\zeta^{(t)}(u)du,$$

where $z = (h-n)/(\sigma\sqrt{N})$. Using (2.3.15), the difference

$$R = 2\pi\left[d^{-1}\sigma\sqrt{N}\,\mathbf{P}\left\{\zeta_N^{(t)} = N + h\right\} - (2\pi)^{-1/2}e^{-z^2/2}\right]$$

can be rewritten as the sum of four integrals $R = I_1 + I_2 + I_3 + I_4$, where

$$\begin{aligned} I_1 &= \int\limits_{-A}^{A} e^{-izu}\left[\varphi_\zeta^{(t)}(u) - e^{-u^2/2}\right]du, \\ I_2 &= \int\limits_{A<|u|\leqslant\varepsilon\sigma\sqrt{N}} e^{-izu}\varphi_\zeta^{(t)}(u)du, \\ I_3 &= \int\limits_{\varepsilon\sigma\sqrt{N}<|u|\leqslant\pi\sigma\sqrt{N}} e^{-izu}\varphi_\zeta^{(t)}(u)du, \\ I_4 &= -\int\limits_{A<|u|} \exp\{-izu - u^2/2\}du; \end{aligned} \tag{21}$$

the positive constants A and ε will be chosen later. Below we will show that by choosing sufficiently large N, n, A and sufficiently small ε the difference R can be made arbitrarily small. To this end we estimate the integrals $I_1 - I_4$.

It is not hard to see that under the hypotheses of Lemma 4 the conditions of Lemma 3 are valid too; therefore $I_1 \to 0$. It is clear also that

$$|I_4| \leqslant \int\limits_{A<|u|} e^{-u^2/2}du$$

and the last expression can be made arbitrarily small by choosing a sufficiently large A.

To estimate integrals I_2 and I_3 we note that (7), (8), (10) and the condition $N\,\mathbf{P}\{\mu^{(1)}(t) > 0\} \leqslant C_2$ imply that

$$\left|\varphi_\zeta^{(t)}(u)\right| \leqslant \left|\varphi\left(\frac{u}{\sigma\sqrt{N}}\right)\right|^N \left(\frac{1 + \varphi^{-1}(u/(\sigma\sqrt{N}))/N}{1 + C_3/N}\right)^N, \tag{22}$$

where C_3 is some positive constant. From the estimations of the function $\varphi(u/(\sigma\sqrt{N}))$ that we obtained in Lemma 2.3.2 it follows that in the integration domains of I_2 and I_3 the inequality

$$|\varphi^{-1}(u/(\sigma\sqrt{N}))| \leqslant C_4 < \infty$$

holds. Therefore, from (22) we get

$$\left|\varphi_\zeta^{(t)}(u)\right| \leqslant C_5 \left|\varphi(u/(\sigma\sqrt{N}))\right|^N,$$

where $C_5 > 0$. From this we see that all reasonings of the proof of Lemma 2.3.2 about I_2 and I_3 tendency to zero are true. Lemma 4 is proved.

Remark 2. Under the hypotheses of Lemma 4 and the condition $\lambda \to 0$ the assertion of Lemma 4 holds for $\zeta_N^{(t+1)}$ because the proof is true without modifications if we take into account Remark 1. If in addition $n^2/N \to \infty$ then the assertion of Lemma 4 holds for $\zeta_N^{(t-1)}$. Indeed by (7), (8), (10) and Theorem 1.3.2

$$\left|\varphi_\zeta^{(t-1)}(u)\right| = \left|\varphi\left(\frac{u}{\sigma\sqrt{N}}\right)\right|^N \left(1 + O\left(\frac{1}{n}\right)\right)^N. \tag{23}$$

The estimations of $\varphi(u)$ from the proof of Lemma 2.3.2 imply that for sufficiently large N, n

$$\left|\varphi\left(\frac{u}{\sigma\sqrt{N}}\right)\right|^N \left(1 + O\left(\frac{1}{n}\right)\right)^N \leqslant \left|\varphi\left(\frac{u}{\sigma\sqrt{N}}\right)\right|^{N/2}$$

and using (23) we see that in this case the proof of Lemma 4 also holds.

3. The limit behaviour of auxiliary probabilities in the critical case

To prove Theorems 1.4 and 1.5 in Section 4 we will use the relation (1.3). In this Section we obtain the asymptotics of the probabilities $\mathsf{P}\{\mu(t) > 0\}$ and $\mathsf{P}\{\nu_N = N + n|\ \mu(t) > 0\}$ from the right-hand side of (1.3) for $\lambda = 1$.

Lemma 1. *Let $t \to \infty$. Then*

$$\mathsf{P}\{\mu(t) > 0\} = 1 - \exp\{-2N/(Bt)\} + o(N/t)$$

uniformly in the integers N such that $N/t \leqslant C < \infty$.

Proof. By Theorem 1.3.3 $\mathsf{P}\{\mu^{(1)}(t) > 0\} = (2/(Bt))(1 + o(1))$. To conclude the proof, it remains to note that

$$\mathsf{P}\{\mu(t) > 0\} = 1 - \mathsf{P}\{\mu(t) = 0\} = 1 - \left(1 - \mathsf{P}\{\mu^{(1)}(t) > 0\}\right)^N.$$

We consider the random variable ν_N under the condition $\mu(t) > 0$. For this variable we denote by $\varphi_t(u)$ the characteristic function. It is not hard to see that

$$\varphi_t(u) = \Delta_{N,t}\left(e^{iu}\right) / \mathsf{P}\{\mu(t) > 0\}, \tag{1}$$

where $\Delta_{N,t}$ is given by (2.6).

Lemma 2. *Let $n, t \to \infty$ in such a way that $n/N^2 \to \infty$, $t(B/n)^{1/2} \to x$, where x is a positive constant. Then for any fixed u*

$$\varphi_t\left(\frac{u}{n}\right) \to \frac{x\sqrt{-2iu}\exp\{-x\sqrt{-2iu}\}}{1 - \exp\{-x\sqrt{-2iu}\}}.$$

Proof. From (2.6) it follows that

$$\Delta_{N,t}\left(e^{iu/n}\right) = f_1^N\left(e^{iu/n}\right) - f_{1,t}^N\left(e^{iu/n}\right); \tag{2}$$

therefore

$$\Delta_{N,t}\left(e^{iu/n}\right) = \Delta_{1,t}\left(e^{iu/n}\right)\sum_{k=1}^{N} f_1^{N-k}\left(e^{iu/n}\right) f_{1,t}^{k-1}\left(e^{iu/n}\right), \tag{3}$$

$$f_{1,t}(e^{iu}) = f_1\left(e^{iu/n}\right) - \Delta_{1,t}\left(e^{iu/n}\right). \tag{4}$$

By analogy with the proof of Lemma 1.3.2 we can easily obtain

$$f_{1,t+1}(w) = wF(f_{1,t}(w)). \tag{5}$$

Since $F''(1) = B$ we have

$$F(w) = F(w_0) + F'(w_0)(w - w_0) + (F''(w_0) + \varepsilon_1(w, w_0))(w - w_0)^2/2,$$

where $\varepsilon_1(w, w_0) \to 0$ for $w \to w_0$, $|w| \leqslant 1$, $|w_0| \leqslant 1$. Note that the last relation is valid uniformly in w_0. Therefore by (2), (5), (1.3.35)

$$\begin{aligned}\Delta_{1,t}(w) &= w(F(f_1(w)) - F(f_{1,t-1}(w))) \\ &= w\Delta_{1,t-1}(w)F'(f_1(w)) - w(F''(f_1(w)) + \varepsilon_2(t, w))\Delta_{1,t-1}^2(w)/2,\end{aligned} \tag{6}$$

where $\varepsilon_2(t, w) \to 0$ for $t \to \infty$ uniformly in $|w| \leqslant 1$. Let W_t be the set of points w, $|w| \leqslant 1$ such that $\Delta_{1,t}(w) = 0$. From (6) we get $W_{t-1} \subseteq W_t$ and for $w \notin W_t$

$$|\Delta_{1,t}(w)/\Delta_{1,t-1}(w) - wF'(f_1(w))| \leqslant C\Delta_{1,t-1}(1), \tag{7}$$

where C is some positive constant, $\Delta_{1,t-1}(1) = \mathbf{P}\{\mu^{(1)}(t-1) > 0\} \to 0$ for $t \to \infty$. In the critical case, $F'(1) = 1$; therefore $\Delta_{1,t}(w)/\Delta_{1,t-1}(w) \to 1$ and beginning with some t_0 the sets of zeros of the functions $\Delta_{1,t}(w)$ and $\Delta_{1,t-1}(w)$ are coincident. By the obvious relation $\Delta_{1,t}(w) \to \mathbf{P}\{\mu^{(1)}(t) > 0\} > 0$ which is true for $w \to 1$ and any fixed t we obtain that in the circle $|w| \leqslant 1$ there exists such neighbourhood of the unit that $|\Delta_{1,t}(w)| > 0$ for $t \leqslant t_0$. This means that the last inequality holds for any t. By (7) in this neighbourhood

$$\Delta_{1,t-1}(w) = \Delta_{1,t}(w)\left(wF'(f_1(w))\right)^{-1}(1 + \varepsilon_2(t, w)), \tag{8}$$

where $\varepsilon_2(t, w) \to 0$ for $t \to \infty$ uniformly in w. Using (6) and (8) we obtain

$$\begin{aligned}\Delta_{1,t}(w) = wF'(f_1(w))\Delta_{1,t-1}(w) - (F''(f_1(w))(2F'(f_1(w)))^{-1} \\ +\varepsilon(t, w))\Delta_{1,t}(w)\Delta_{1,t-1}(w),\end{aligned} \tag{9}$$

where $\varepsilon(t, w) \to 0$ for $t \to \infty$ uniformly in w in the considered neighbourhood. Dividing both sides of (9) by $\Delta_{1,t}(w)\Delta_{1,t-1}(w)$ we get

$$h_{t-1}(w) = a(w)h_t(w) - b(w) - \varepsilon(t, w), \tag{10}$$

where

$$\begin{aligned} h_t(w) &= (\Delta_{1,t}(w))^{-1}, \\ a(w) &= wF'(f_1(w)), \\ b(w) &= F''(f_1(w))(2F'(f_1(w)))^{-1}. \end{aligned}$$

Let

$$H(z) = \sum_{t=0}^{\infty} h_t(w) z^t.$$

Taking into account the equality $\Delta_{1,0}(w) = f_1(w)$ we find from (10) that

$$H(z) = \frac{1}{f_1(w)(1 - z/a(w))} + \frac{zb(w)}{a(w)(1-z)(1-z/a(w))}$$
$$+(a(w)(1 - z/a(w)))^{-1} \sum_{t=1}^{\infty} \varepsilon(t, w) z^t.$$

From this relation we can deduce that

$$h_t(w) = \frac{1}{a^t(w)} \left(\frac{1}{f_1(w)} + \frac{b(w)(1 - a^t(w))}{1 - a(w)} \sum_{k=0}^{t} \varepsilon(k, w) a^k(w) \right). \tag{11}$$

Let $n, t \to \infty$, $t(B/n)^{1/2} \to x$ and $z = e^{iu/n}$. Using Lemma 1.3.3 we obtain

$$\begin{aligned} a(w) &= 1 - \sqrt{-2iBu/n}(1 + o(1)), \\ a^t(w) &= \exp\{-x\sqrt{-2iu}\}(1 + o(1)), \\ b(w) &= B/2(1 + o(1)). \end{aligned} \tag{12}$$

Since $\varepsilon(t, z)$ uniformly tends to zero we find that $|\varepsilon(k, w)| \leqslant \varepsilon(k)$ and $\varepsilon(k) \to 0$ for $k \to \infty$. Also, $|a(w)| = |wF'(w)| \leqslant F'(1) = 1$; therefore

$$\left| \sum_{k=0}^{t} \varepsilon(k, w) a^k(w) \right| \leqslant \sum_{k=0}^{t} \varepsilon(k) = o(t).$$

This together with (11), (12) implies that

$$\Delta_{1,t}\left(e^{iu/n}\right) = \frac{2}{Bt} \frac{x\sqrt{-2iu}\exp\{-x\sqrt{-2iu}\}}{1 - \exp\{-x\sqrt{-2iu}\}}(1 + o(1)). \tag{13}$$

By Lemma 1.3.3

$$f_1\left(e^{iu/n}\right) = 1 - \sqrt{-2iu/(Bn)}(1 + o(1)); \tag{14}$$

therefore, from (4) and (13) we obtain

$$\sum_{k=1}^{N} f_1^{N-k}\left(e^{iu/n}\right) f_{1,t}^{k-1}\left(e^{iu/n}\right) = N(1 + o(1)). \tag{15}$$

Combining Lemma 1, (1), (3), (13) and (15) we get the assertion of Lemma 2.

Lemma 3. *Let the hypotheses of Lemma 2 hold. Then the distribution of ν_N/n under the condition $\mu(t) > 0$ converges weakly to the distribution with the characteristic function*

$$\Psi(u) = \frac{x\sqrt{-2iu}\exp\{-x\sqrt{-2iu}\}}{1 - \exp\{-x\sqrt{-2iu}\}}$$

and density

$$g(y) = \frac{x}{y\sqrt{2\pi y}} \sum_{k=1}^{\infty} \left(\frac{k^2 x^2}{y} - 1 \right) \exp\left\{ -\frac{k^2 x^2}{2y} \right\}, \quad y > 0. \tag{16}$$

Proof. In Lemma 2 we saw that the characteristic function $\varphi_t(u/n)$ tends to

$$\Psi(u) = \frac{x\sqrt{-2iu}\exp\{-x\sqrt{-2iu}\}}{1-\exp\{-x\sqrt{-2iu}\}} = \sum_{k=1}^{\infty}\Psi_k(u), \quad u \neq 0,$$

where

$$\Psi_k(u) = x\sqrt{-2iu}\exp\{-kx\sqrt{-2iu}\}, \quad k = 1, 2, \ldots$$

Note that $\Psi_k(u)$ is the Fourier transformation of the function

$$g_k(y) = \frac{x}{y\sqrt{2\pi y}}\left(\frac{k^2x^2}{y} - 1\right)\exp\left\{-\frac{k^2x^2}{2y}\right\}, \quad y > 0.$$

This means that the characteristic function $\varphi(u)$ corresponds to the density $g(y)$.

Lemma 4. *Let $y = (N + h)/n$. Under the conditions of Lemma 2 for non-negative integers k which are divided by d*

$$n\,\mathbf{P}\{\nu_N/n = y|\ \mu(t) > 0\} = dg(y)(1 + o(1))$$

uniformly in y lying in any finite interval.

Proof. In Lemma 3 we saw that the density $g(y)$ corresponds to the characteristic function $\Psi(u)$. Since $\Psi(u)$ is absolutely integrable we can obtain by inversion formula

$$g(y) = (2\pi)^{-1}\int_{-\infty}^{\infty} e^{-iuy}\Psi(u)du,$$

$$\mathbf{P}\{\nu_N = N + h|\ \mu(t) > 0\} = d(2\pi n)^{-1}\int_{-\pi n/d}^{\pi n/d} e^{-iuy}\varphi_t(u/n)du.$$

Let us prove that the difference

$$R(y) = 2\pi n d^{-1}\,\mathbf{P}\{\nu_N = N + h|\ \mu(t) > 0\} - 2\pi g(y)$$

tends to zero uniformly in y. We represent this difference as the sum $R(y) = I_1 + I_2 + I_3 + I_4$, where

$$I_1 = \int_{|u|\leqslant A} e^{-iuy}(\varphi_t(u/n) - \Psi(u))du,$$

$$I_2 = -\int_{|u|>A} e^{-iuy}\Psi(u)du,$$

$$I_3 = \int_{A<|u|\leqslant\varepsilon n} e^{-iuy}\varphi_t(u/n)du,$$

$$I_4 = \int_{\varepsilon n\leqslant|u|\leqslant\pi n/d} e^{-iuy}\varphi_t(u/n)du$$

and we show that each of these integrals can be made arbitrarily small by suitably choosing N, n, A, ε.

By Lemma 2, $I_1 \to 0$ for any fixed A. The function $\Psi(u)$ is absolutely integrable; therefore the quantity $|I_2|$ can be made arbitrarily small by choosing A.

From (1), (3), (15) and Lemma 1 it follows that for sufficiently large n, t there exists such positive constant C_1 that for $|u/n| \leqslant \varepsilon$

$$|\varphi_t(u/n)| \leqslant C_1 \left|\Delta_{1,t}\left(e^{iu/n}\right)\right| t. \tag{17}$$

By Lemma 1.3.3 as $u \to 0$

$$\left|f_1\left(e^{iu}\right)\right| = \left|1 - \sqrt{2(1-e^{iu})/B}(1+o(1))\right| = 1 - \sqrt{|u|/B}(1+o(1))$$

and there exist positive constants C_2 and ε such that for $|u| \leqslant \varepsilon$

$$\left|f_1\left(e^{iu}\right)\right| \leqslant 1 - C_2\sqrt{|u|}. \tag{18}$$

For any natural k at $|w| < 1$

$$|\Delta_{1,k}(w)| \leqslant \Delta_{1,k}(1) = \mathbf{P}\{\mu^{(1)}(k) > 0\}$$

and by Theorem 1.3.3 for $k \to \infty$

$$\mathbf{P}\{\mu^{(1)}(k) > 0\} = 2(Bk)^{-1}(1+o(1));$$

therefore there exists such $C_3 > 0$ that

$$|\Delta_{1,k}(w)| \leqslant C_3 k^{-1}. \tag{19}$$

From (6) it follows that

$$\Delta_{1,t}(w) = w(F(f_1(w)) - F(f_{1,t-1}(w)))$$
$$= w\Delta_{1,t-1}(w)\sum_{k=1}^{\infty} p_k \sum_{l=1}^{k} f_{1,t-1}^{l-1}(w) f_1^{k-l}(w).$$

Since the process is critical from the last equality and relations $|f_1(w)| \leqslant 1$, $|f_{1,t}(w)| \leqslant 1$, $|w| \leqslant 1$ we find that

$$\begin{aligned}|\Delta_{1,t}(w)| &\leqslant |\Delta_{1,t-1}(w)| \sum_{k=1}^{\infty} p_k(k|f_1(w)| + (1 - |f_1(w)|)) \\ &= |\Delta_{1,t-1}(w)|(|f_1(w)| + (1-p_0)(1-|f_1(w)|)) \\ &= |\Delta_{1,t-1}(w)|(1 - p_0(1-|f_1(w)|)).\end{aligned}$$

From this we obtain that for $k = 0, 1, \ldots, t$, $|w| \leqslant 1$

$$|\Delta_{1,t}(w)| \leqslant |\Delta_{1,k}(w)|(1 - p_0(1-|f_1(w)|))^{t-k}. \tag{20}$$

Using (18)–(20) we find that there exists such $\varepsilon > 0$ that for $|u| \leqslant \varepsilon$

$$\begin{aligned}\left|\Delta_{1,t}\left(e^{iu}\right)\right| &\leqslant C_3 k^{-1}(1 - C_2 p_0\sqrt{|u|})^{t-k} \\ &\leqslant C_3 k^{-1} \exp\left\{-C_2 p_0 \sqrt{|u|}(t-k)\right\}.\end{aligned}$$

This implies that for some positive C_4, C_5 for $|u| \leqslant \varepsilon$

$$\left|\Delta_{1,t}\left(e^{iu}\right)\right| \leqslant C_4 t^{-1} \exp\left\{-C_5 t\sqrt{|u|}\right\}. \tag{21}$$

From this and (17) we obtain that at $|u/n| \leqslant \varepsilon$

$$|\varphi_t(u/n)| \leqslant C_6 \exp\left\{-C_7\sqrt{|u|}\right\},$$

where C_6, C_7 are positive constants. From this we see that

$$|I_3| \leqslant \int\limits_{A<|u|\leqslant \varepsilon n} |\varphi_t(u/n)|du \leqslant 2C_6 \int\limits_{A<u} \exp\{-C_7\sqrt{u}\}du$$

and $|I_3|$ can be made arbitrarily small by choosing A.

Since $\Delta_{1,0}(w) = f_1(w)$, the estimation (20) at $k=0$ gives us the inequality

$$|\Delta_{1,t}(w)| \leqslant (1 - p_0(1 - |f_1(w)|))^t. \tag{22}$$

The maximum span of the distribution ν_N is equal to d; therefore for any $\varepsilon > 0$ there exists such $q_1 < 1$ that for $\varepsilon \leqslant |u| \leqslant \pi/d$ the estimation $|f_1(e^{iu})| \leqslant q_1$ holds. From this and (22) we obtain that for these u there exists such $q < 1$ that

$$\left|\Delta_{1,t}\left(e^{iu}\right)\right| \leqslant q^t. \tag{23}$$

Therefore, from (17) we get

$$|I_4| \leqslant \int\limits_{\varepsilon n\leqslant|u|\leqslant\pi n/d} |\varphi_t(u/n)|du \leqslant 2\pi n t q^t \to 0. \tag{24}$$

Lemma 4 is proved.

Lemma 5. *Let $n, t \to \infty$ in such a way that $N/\sqrt{Bn} \to z$, $t(B/n)^{1/2} \to x$, where x is a positive constant. Then for any fixed u*

$$\varphi_t(u/n) \to \Psi_1(u)$$

uniformly in z lying in any finite interval, where

$$\Psi_1(u) = \frac{\exp\{-z\sqrt{-2iu}\}\exp\{-zf(x,u)/x\}}{1-\exp\{-2z/x\}}$$

and function $f(x,u)$ is given by (1.8).

Proof. By (14)

$$f_1^N\left(e^{iu/n}\right) = \exp\left\{-z\sqrt{-2iu}\right\}(1+o(1))$$

and from (2)–(4), (13)–(15) it is not hard to obtain

$$f_{1,t}^N\left(e^{iu/n}\right) = \exp\{-zf(x,u)/x\}(1+o(1)).$$

Therefore, from (2) we see

$$\Delta_{N,t}\left(e^{iu/n}\right) = \exp\left\{-z\sqrt{-2iu}\right\} - \exp\{-zf(x,u)/x\} + o(1).$$

This, (1) and Theorem 1.3.3 imply the assertion of Lemma 5.

Lemma 6. *Under the conditions of Lemma* 5 *for non-negative integers k which are divided by d*

$$\mathbf{P}\{\nu_N/n = y|\ \mu(t) > 0\} = \frac{d(1+o(1))}{2\pi n(1-\exp\{-2z/x\})}$$

$$\times \int\limits_{-\infty}^{\infty} e^{-iuy}\left(\exp\{-z\sqrt{-2iu}\} - \exp\{-zf(x,u)/u\}\right)du$$

uniformly in z and $y = (N+h)/n$ lying in any domain in the form $0 < y_0 \leqslant z, y \leqslant y_1 < \infty$, where the function $f(x,u)$ is given by (1.8).

Proof. In Lemma 5 we saw that the characteristic function $\varphi_t(u/n)$ tends to $\Psi_1(u)$, which corresponds to the density

$$g_1(y) = (2\pi)^{-1} \int_{-\infty}^{\infty} e^{-iuy}\Psi_1(u)du.$$

Besides

$$\mathbf{P}\{\nu_N = N+h |\, \mu(t) > 0\} = d(2\pi n)^{-1} \int_{-\pi n/d}^{\pi n/d} e^{-iuy}\varphi_t(u/n)du.$$

Let us consider that

$$R(y) = 2\pi n d^{-1}\,\mathbf{P}\{\nu_N = N+h |\, \mu(t) > 0\} - 2\pi g_1(y)$$

tends to zero. As usually we represent this difference as $R(y) = I_1 + I_2 + I_3 + I_4$, where

$$I_1 = \int_{|u|\leqslant A} e^{-iuy}(\varphi_t(u/n) - \Psi_1(u))du,$$

$$I_2 = -\int_{|u|>A} e^{-iuy}\Psi_1(u)du,$$

$$I_3 = \int_{A<|u|\leqslant \varepsilon n} e^{-iuy}\varphi_t(u/n)du,$$

$$I_4 = \int_{\varepsilon n<|u|\leqslant \pi n/d} e^{-iuy}\varphi_t(u/n)du;$$

A and ε are defined below.

By Lemma 5 $I_1 \to 0$. The quantity $\Psi_1(u)$ is absolutely integrable; therefore $|I_2|$ can be made arbitrarily small by choosing A.

By (3)

$$\left|\Delta_{N,t}\left(e^{iu/n}\right)\right| \leqslant N\left|\Delta_{1,t}\left(e^{iu/n}\right)\right|; \tag{25}$$

hence (21) implies that

$$\left|\Delta_{N,t}\left(e^{iu/n}\right)\right| \leqslant C_1 \exp\left\{-C_2\sqrt{|u|}\right\},$$

where C_1, C_2 are some positive constants. From this, (1) and Theorem 1.3.3 we obtain

$$|\varphi_t(u/n)| \leqslant C_3 \exp\{-C_2\sqrt{|u|}\}, \quad C_3 > 0;$$

therefore

$$|I_3| \leqslant C_3 \int_{u>A} \exp\{-C_2\sqrt{u}\}du.$$

Obviously, this integral can be made arbitrarily small by choosing A. Finally, using (1), (23) and (25) we find the relation (24) for the integral I_4.

4. Proofs of the main results

Under the conditions of Theorem 1.1 we obtain from Lemmas 2.4, 2.3.2 and Theorem 1.3.2 that

$$\mathsf{P}\left\{\zeta_N^{(t+1)} = N+n\right\} / \mathsf{P}\{\nu_N = N+n\} \to 1. \tag{1}$$

Therefore (1.2) and Lemma 2.1 imply that

$$\mathsf{P}\{\tau < t+1\} = \exp\left\{-Nm^{t+1}\right\} + o(1). \tag{2}$$

Since $Nm^{t+2} \to 0$ it follows from Remark 2.2 and Lemmas 2.3.2, 2.1 by analogy with (2) that

$$\mathsf{P}\{\tau < t+2\} = 1 + o(1), \quad \mathsf{P}\{\tau < t\} = o(1).$$

From this and (2) we get the assertion of Theorem 1.1.

To prove Theorem 1.2 we note that the condition $Nm^t \to \beta$ and Theorem 1.3.2 imply the inequality $N\,\mathsf{P}\{\mu(t+k) > 0\} \leqslant C_2 < \infty$. Using Lemmas 2.4 and 2.3.2 we obtain the relation (1) and to conclude the proof of Theorem 1.2 it remains to use (1.2) and the first assertion of Lemma 2.1.

Let the hypotheses of Theorem 1.3 hold and $t = t(N,n)$ be given by (2.1). Using Theorem 1.3.4 it is not hard to check that in this case the conditions of Lemma 2.4 are true. Therefore, from Lemma 2.3.2 follows the relation (1); hence (1.2) and the second assertion of Lemma 2.1 imply Theorem 1.3.

Let $n \to \infty$ in such a way that n takes values which are divided by d, $N/\sqrt{Bn} \to z$ and $t = x\sqrt{n/B}$, where z, x are positive constants. Using (1.3) and Lemmas 2.3.3, 3.1, 3.6 we obtain

$$\begin{aligned} \mathsf{P}\left\{(B/n)^{1/2}\tau < x\right\} &= 1 - \left(z\sqrt{2\pi}\right)^{-1} e^{z^2/2} \\ &\times \int_{-\infty}^{\infty} e^{-iu}\left(\exp\left\{-z\sqrt{-2iu}\right\} - \exp\{-zf(x,u)/x\}\right) du(1+o(1)). \end{aligned} \tag{3}$$

Since $\exp\{-\sqrt{-2iu}\}$ is the characteristic function of the distribution with the density $(2\pi y^3 e^{1/y})^{-1/2}$ it follows that

$$(z\sqrt{2\pi})^{-1} e^{z^2/2} \int_{-\infty}^{\infty} \exp\left\{-iu - z\sqrt{-2iu}\right\} du = 1.$$

From this and (3) we get the assertion of Theorem 1.4.

If $n \to \infty$, $n/N^2 \to \infty$, then putting $t = x\sqrt{n/B}$, where x is a positive constant, we obtain Theorem 1.5 from (1.3) and Lemmas 2.3.4, 3.1, 3.4.

We can find some sufficient conditions for the series $t(N,n)$ which are defined in Theorem 1.1. Let $g(N,n,s) = \ln N + s\ln m$ and $t = [-(\ln N)/(\ln m)]$ where square brackets denote the integer part. Since

$$Nm^t = \exp\{g(N,n,t)\}$$

it is not hard to obtain (by analogy with Theorem 2.6.1) the next assertion.

Theorem 1. *Let n be divided by d and $n, N \to \infty$ in such a way that $n/N \to 0$. Let $m = n/(N+n)$ and one of the conditions:*

1) $g(N, n, t) \to \infty$;

2) $g(N, n, t)$ *is limited;*

3) *there exists a series* $t = t(N, n)$ *such that* $g(N, n, t)$ *is limited;*

hold. Then $Nm^s \to \infty$, Nm^{s+1} *is limited, where under the condition* 1) $s = t$ *and under the conditions* 2), 3) $s = t - 1$.

5. Additions and references

The methods of studying the limit behaviour of the random forest height are a natural continuation of the methods proposed by V. F. Kolchin for the study of the random tree height and described in most detail in [38]. Limit theorems for the height of a random forest were first obtained in [57], where the forest class $\mathfrak{F}'_{N,n}$ (see Example 1.1.1) with the uniform distribution set on it was considered. These results readily follow from Theorems 1.1–1.5 in which it is necessary to put $d = 1$, $B = 1$, $j = 1$, $l = 0$, and the Poisson distribution generating function (1.3.52) should be used instead of $F_\lambda(z)$ in Theorem 1.2. Note also that as $n/N \to 0$, the result of Theorem 1.1 holds for forests from $\mathfrak{F}'_{N,n}$ even in the absence of the restriction $n^2/N \to \infty$ (see below). The height of a random forest from the set $\mathfrak{F}'_{N,n}(R)$ described in Example 1.1.2 was investigated by I. B. Kalugin in the papers [31, 32] for two cases: $N = cn(1 + o(1))$, where $c \in (0, 1)$ and $N = o(\sqrt{n})$. Limit distributions of the height of a random forest consisting of plane planted trees (Example 1.1.3) were obtained in [63]. Just as in the case of a forest from $\mathfrak{F}'_{N,n}$, they follow from Theorems 1.1–1.5, if $d = j = 1$, $l = 0$, $B = 2$ and the generating function of the geometric distribution (1.3.43) is used instead of $F_\lambda(z)$. The generalization of these results for the class $\mathfrak{F}_{N,n}$ suggested in Chapter 4 is based on the results presented in [64].

Note that the case $n^2/N \leqslant C < \infty$ is not considered in theorems of Section 1. Let $n^2/N \to 0$. It is clear from (3.1.5), Lemma 2.2.1 and Theorem 2.1.1 that the maximum tree size in this case is asymptotically not over $j + 1$. This means that the forest height is asymptotically equal to one; hence $\mathbf{P}\{\tau = 1\} = 1 + o(1)$. If $0 < C_1 \leqslant n^2/N \leqslant C_2 < \infty$, it then follows from analogous reasoning that the forest height is concentrated on the set $\{1, 2\}$. From the equality (1.3.56) and Lemma 1.3.4 get that

$$\mathbf{P}\{\tau < 2\} = \mathbf{P}\{\mu(2) = 0 |\, \nu_N = N + n\}$$

$$= \mathbf{P}\left\{\mu^{(1)}(1) + \cdots + \mu^{(N)}(1) = n,\ \mu(2) = 0\right\} / \mathbf{P}\{\nu_N = N + n\}$$

$$= \frac{N+n}{N} \frac{\mathbf{P}\{\xi_1 + \cdots + \xi_N = n\}(\mathbf{P}\{\xi_1 = 0\})^n}{\mathbf{P}\{\xi_1 + \cdots + \xi_{N+n} = n\}},$$

where $\xi_1, \ldots, \xi_{N+n}$ are independent identically distributed random variables with the distribution (1.4). Hence

$$\mathbf{P}\{\tau < 2\} = \frac{N+n}{N}(p_0(\lambda)F(\lambda))^n \frac{\sum_1 p_{k_1} \cdots p_{k_N}}{\sum_2 p_{k_1} \cdots p_{k_{N+n}}}, \tag{1}$$

where summation in $\sum_1$ is carried out along non-negative integers $k_1, \ldots, k_N$, such that $k_1 + \cdots + k_N = n$, and in $\sum_2$ along non-negative integers $k_1, \ldots, k_{N+n}$, such that $k_1 + \cdots + k_{N+n} = n$. Formula (1) can easily be used to obtain the limit distributions of the height of random forests of different types. Namely, for the class of forests

$\mathfrak{F}'_{N,n}$ on which uniform distribution was set, the ratio (1.4) is the Poisson distribution; hence

$$\sum\nolimits_1 \frac{n!}{k_1!\dots k_N!} = N^n, \quad \sum\nolimits_2 \frac{n!}{k_1!\dots k_{N+n}!} = (N+n)^n$$

and using (1) we find that for $\mathfrak{F}'_{N,n}$

$$\begin{aligned} \mathsf{P}\{\tau = 1\} &= \mathsf{P}\{\tau < 2\} = \exp\{-n^2/N\} + o(1), \\ \mathsf{P}\{\tau = 2\} &= 1 - \mathsf{P}\{\tau = 1\} = 1 - \exp\{-n^2/N\} + o(1). \end{aligned} \tag{2}$$

A similar argument can be applied to the class of forests $\mathfrak{F}''_{N,n}$ with the geometric distribution corresponding to it, as seen from Example 1.3.1. Using (1) and the evident equalities

$$\sum\nolimits_1 2^{-(N+n)} = 2^{-(N+n)} \binom{N+n-1}{n}$$

$$\sum\nolimits_2 2^{-(N+2n)} = 2^{-(N+2n)} \binom{N+2n-1}{n}$$

we get that the relations (2) hold for the forests from $\mathfrak{F}''_{N,n}$ as well.

Lemma 2.2 is analogous to Lemma 2.2.3 from [38]. The results of Section 3 are a generalization of theorems and lemmas from Section 2.4 [38].

Relying on the methods used to prove the theorem about the limit behaviour of the random forest height one can obtain results concerning the distributions of the number of vertices in the strata of a random forest. The paper [84] deals with random trees with labelled vertices and uniform probability distribution. In the paper, the limit distribution of the number of vertices in the t-th stratum of a random tree as $n, t \to \infty$, $t/\sqrt{n} \to \beta$, where β is a positive constant was obtained. In [38], the result was extended to the lower strata of a random tree, which satisfy the condition $t/\sqrt{n} \to 0$ as $n \to \infty$. Limit distributions of the number of vertices in strata of a random forest from $\mathfrak{F}'_{N,n}$, $\mathfrak{F}'_{N,n}(R)$ were obtained in [31, 32] for two cases of N and n tending to infinity: $t/\sqrt{n} \to 0$, $N = z\sqrt{n}$, where z is a positive constant, and $N = n(1-\delta)/\delta$, $n\delta^t \to \infty$, where $0 < \delta < 1$. For the set $\mathfrak{F}'_{N,n}$ these results were extended to the case $N/\sqrt{n} \leqslant \gamma < \infty$, $t/\sqrt{n} \leqslant \beta < \infty$ in [58, 60]. It follows from Theorems 1.1–1.5 that the height of a random forest (maximum stratum number) under these restrictions has the order of $\sqrt{n}$. Therefore, the results obtained in [58, 60] provide a comprehensive picture of the limit behaviour of the number of vertices in the strata of a forest from $\mathfrak{F}'_{N,n}$. Similar results for forests from $\mathfrak{F}''_{N,n}$ were demonstrated in [20].

For the forests from $\mathfrak{F}_{N,n}$ considered in the present book, let $\mu(t, \mathfrak{F})$ denote the random variable equal to the number of vertices in the t-th stratum; other notations are introduced in Section 1. Using the methods suggested in [32] we get the following statements (see [12, 69]).

Theorem 1. *Let $N, n \to \infty$ in such a way that $n/N \leqslant C_1 < \infty$, $Nm^t \to \infty$. Let $t \geqslant 3$ for $n/N \to 0$ and $t \to \infty$ for $n/N \geqslant C_2 > 0$. Then*

$$\mathsf{P}\{\mu(t, \mathfrak{F}) = k\} = \frac{d(1+o(1))}{\sqrt{2\pi(N+n)B_\lambda m^{t-1}}} e^{-y^2/2}$$

uniformly in $y = (k - Nm^t)/\sqrt{(N+n)B_\lambda m^{t-1}}$ *lying in any finite interval.*

Theorem 2. *Let* $N, n, t \to \infty$ *in such a way that* $n/N \to \infty$, $n/N^2 \to 0$, $tN/(n\ln(N^2/n)) \to 0$, $tN/n \to \infty$ *and the inequality* $\max\{n/N, N/\sqrt{n}\} \leqslant N^L m^{tL/2} n^{-L/2}$ *be valid for some positive* L *for sufficiently large* N, n. *Then*

$$\mathbf{P}\{\mu(t,\mathfrak{F}) = k\} = \frac{d(1+o(1))}{\sqrt{2\pi Bnm^t}} e^{-y^2/2}$$

uniformly in $y = (k - Nm^t)/\sqrt{Bnm^t}$ *lying in any finite interval.*

Theorem 3. *Let* $N, n, t \to \infty$ *in such a way that* $n/N \to \infty$, $n/N^2 \to 0$, $tN/n \to \alpha$, $0 < \alpha < \infty$ *and the inequality* $n/N < (N/\sqrt{n})^L$ *be valid for some positive* L *for sufficiently large* N, n. *Then*

$$\mathbf{P}\{\mu(t,\mathfrak{F}) = k\} = \frac{d(1+o(1)}{\sqrt{2\pi HBn}} e^{-y^2/2}$$

uniformly in $y = (k - Nm^t)/\sqrt{HBn}$ *lying in any finite interval, where*

$$H = e^{-2\alpha}(1 - e^{-\alpha})(1 + Q(1-\rho^2)e^{-\alpha}),$$

$$Q = 1 - 2\alpha e^{-\alpha}, \quad \rho = \frac{(\alpha - 1 + e^{-\alpha})e^{-\alpha/2}}{\sqrt{(1-e^{-\alpha})Q}}.$$

Theorem 4. *Let* $N, n \to \infty$ *in such a way that* $n/N \to \infty$, $n/N^2 \to 0$, $tN/n \to 0$, $N^4t^2/n^3 \geqslant C > 0$ *and the inequality* $n/N < (N/\sqrt{n})^L$ *be valid for some positive* L *for sufficiently large* N, n. *Then*

$$\mathbf{P}\{\mu(t,\mathfrak{F}) = k\} = \frac{d(1+o(1))}{\sqrt{2\pi NBt}} e^{-y^2/2}$$

uniformly in $y = (k - Nm^t)/\sqrt{NBt}$ *lying in any finite interval.*

Theorem 5. *Let* $N, n \to \infty$ *in such a way that* $n/N \to \infty$, $n/N^2 \to 0$, $N^4t^2/n^3 \to 0$. *Then for any fixed* x

$$\mathbf{P}\left\{\frac{\mu(t,\mathfrak{F}) - Nm^t}{\sqrt{BNt}} \leqslant x\right\} \to \frac{1}{\sqrt{2\pi}} \int_{-\infty}^{x} e^{-u^2/2} du.$$

Theorem 6. *Let* $N, n \to \infty$ *in such a way that* $0 < C_1 \leqslant n/N \leqslant C_2 < \infty$. *Then for any fixed* t

$$\mathbf{P}\{\mu(1,\mathfrak{F}) = k_1, \ldots, \mu(t,\mathfrak{F}) = k_t\}$$
$$= (2\pi(N+n)B_\lambda d^{-2})^{-1/2}\sqrt{\det Q(m)}\exp\{-2^{-1}uQ(m)u'\}(1+o(1)),$$

uniformly in $u_i = (k_i - Nm^i)/\sqrt{(N+n)B_\lambda}$, $i = 1, \ldots, t$, *lying in any finite intervals where* $u = (u_1, \ldots, u_t)$, $Q(m) = ||q_{ij}(m)||_{i,j=1}^t$ *is a matrix with the following elements:*

$$q_{ii}(m) = \begin{cases} ((1-m)^3 + m^{t+1-i}(1+m))/(m^t(1-m)), & 1 \leqslant i \leqslant t-1; \\ 1/(m^t(1-m)), & i = t; \end{cases}$$

$$q_{ij}(m) = q_{ji}(m) = \begin{cases} \dfrac{(1-m)^3 - m^{t+1-i}}{m^t(1-m)}, & 1 \leqslant i \leqslant t-2,\ j = i+1; \\ (1-m)^3/(m^t(1-m)), & 1 \leqslant i \leqslant t-3,\ i+2 \leqslant j \leqslant t-1; \\ (1-m)^2/(m^t(1-m)), & 1 \leqslant i \leqslant t-2,\ j = t; \\ (1-2m)/(m^t(1-m)), & i = t-1, j = t. \end{cases}$$

Theorem 7. *Let $n, t \to \infty$, $n/N^2 \to \infty$, $N/t \to 0$, $t/\sqrt{n} \to 0$. Then for any fixed $x > 0$*

$$\mathbf{P}\left\{\frac{2\mu(t,\mathfrak{F})}{Bt} \leqslant x\right\} \to 1 - e^{-x} - xe^{-x}.$$

Theorem 8. *Let $N, n, t \to \infty$, $n/N^2 \to \infty$, $2N/(Bt) \to \alpha$, where α is a positive constant. Then for any fixed $x > 0$*

$$\mathbf{P}\left\{\frac{2\mu(t,\mathfrak{F})}{Bt} \leqslant x\right\} \to e^{-\alpha} \sum_{k=0}^{\infty} \frac{\alpha^k}{k!} K_{2k+4}(2x),$$

where $K_{2k+4}(2x)$ is the value of the distribution function χ^2 with $2k+4$ degrees of freedom.

Theorem 9. *Let $N, n \to \infty$, $N/\sqrt{n} \leqslant \gamma < \infty$, $N/t \to \infty$, $\mathbf{E}\xi_1^3 < \infty$. Then for any fixed x*

$$\mathbf{P}\left\{\frac{\mu(t,\mathfrak{F})}{\sqrt{NBt}} \leqslant x\right\} \to \frac{1}{\sqrt{2\pi}} \int_{-\infty}^{x} e^{-u^2/2} du.$$

Theorem 10. *Let $n, t \to \infty$, $t\sqrt{B/n} \to \beta$, where β is a positive constant, $n/N^2 \to \infty$, $\mathbf{E}\xi_1^3 < \infty$. Then for any fixed $x_2 > x_1 > 0$*

$$\mathbf{P}\left\{x_1 \leqslant \frac{2\mu(t,\mathfrak{F})}{Bt} \leqslant x_2\right\} \to F(x_1, x_2),$$

where

$$F(x_1, x_2) = \int_{x_1}^{x_2} \int_0^1 \frac{x}{(1-y)^{3/2}} \exp\left\{-\frac{x^2\beta^2}{8(1-y)}\right\} dG_\beta(x, y)$$

and the distribution function $G_\beta(x, y)$ has the characteristic function

$$\Psi(\theta_1, \theta_2) = \left(\frac{\operatorname{sh} u}{u} - i\theta_1 \left(\frac{2\operatorname{sh}(u/2)}{u}\right)^2\right)^{-1}, \quad u = \beta\sqrt{-2i\theta_2}.$$

Theorem 11. *Let $N, n, t \to \infty$, $t\sqrt{B/n} \to \beta$, $N/\sqrt{n} \to \gamma$, where β, γ are positive constants, $\mathbf{E}\xi_1^3 < \infty$. Then for any fixed $x_2 > x_1 > 0$*

$$\mathbf{P}\left\{x_1 \leqslant \frac{2\mu(t,\mathfrak{F})}{Bt} \leqslant x_2\right\} \to F_\gamma(x_1, x_2),$$

where

$$F_\gamma(x_1, x_2) = \int_{x_1}^{x_2} \int_0^1 \frac{\beta x\sqrt{B} \exp\{-x^2\beta^2/(8(1-y))\}(1 - e^{-\gamma\sqrt{B}/\beta})}{2\gamma(1-y)^{3/2}} dG_{\beta,\gamma}(x, y)$$

and the distribution function $G_{\beta,\gamma}(x,y)$ has the characteristic function

$$\Psi_\gamma(\theta_1,\theta_2) = (1 - \exp\{-2\gamma/(\beta\sqrt{B})\})^{-1}$$
$$\times \exp\{-\gamma\sqrt{-2i\theta_2 B}\}(\exp\{-2\gamma f(\theta_1,\theta_2)/(\beta\sqrt{B})\}$$
$$- \exp\{-2\gamma f(\theta_2)/(\beta\sqrt{B})\}),$$
$$f(\theta_2) = ue^{-u}(1-e^{-u})^{-1}, \quad u = \beta\sqrt{-2i\theta_2},$$
$$f(\theta_1,\theta_2) = e^{-u}((-i\theta_1 - u/2)^{-1} + (1-e^{-u})/u)^{-1}.$$

BIBLIOGRAPHY

1. Aldous D. J. Brownian excursions, critical random graphs and the multiplicated coalescent. *Ann. Probab.*, **25**, 1997, 812–854.
2. Aldous D. J. Stochastic Coalescence. *Extra Volume ICM*, III, Berlin, 1998, pp. 205–211.
3. Alexander K. S. Percolation and minimal spaning forests in infinite graphs. *Ann. Probab.*, **23**, 1995, 87–104.
4. Athreya K. B., Ney P. E. *Branching Processes.* Springer, Berlin, 1972.
5. Bollobas B. *Random Graphs.* Academic Press, London, 1985.
6. Borovkov A. A. *Probability Theory.* Nauka, Moscow, 1976 (in Russian).
7. Britikov V. E. Limit theorems for maximum tree size in a random forest from unrooted trees. *Probabilistic problems of discrete mathematics.* MIEM, Moscow, 1987, pp. 84–91 (in Russian).
8. Britikov V. E. Asymptotic number of forests from unrooted trees. *Math. Notes*, **43**, 1988, 387–394.
9. Britikov V. E. The limit behaviour of the number of trees of a given size in a random forest of non-rooted trees. *Probabilistic problems of discrete mathematics.* MIEM, Moscow, 1988, pp. 7–12 (in Russian).
10. Britikov V. E. *Random forests and evolution of random graphs. Candidate dissertation.* MIEM, Moscow, 1990 (in Russian).
11. Britikov V. E. On the random graph structure near the critical point. *Discrete Math. Appl.*, **1**, No. 3, 1991, 301–309.
12. Cheplyukova I. A. Limit distributions of the number of vertices in the layers of a random forest. *Discrete Math. Appl.*, **7**, 1997, 515–522.
13. Cheplyukova I. A. Emergence of the giant tree in a random forest. *Discrete Math. Appl.*, **8**, 1998, 17–34.
14. Cheplyukova I. A., Pavlov Yu. L. Random simply generated forests. *7th Vilnius Conf. on Probab. Theory. 22nd European Meeting of Statisticians. Abstracts.* TEV, Vilnius, 1998, 177.
15. Doyle P. G., Snell J. L. *Random Walks and Electric Networks.* Math. Assoc. of America. Washington, 1984.
16. Drmota M. The height distribution of leaves in rooted trees. *Discrete Math. Appl.*, **4**, 1994, 45–48.
17. Drmota M. On nodes of given degree in random trees. *Probabilistic methods in discrete mathematics. Proc. IV Intern. Petrozavodsk Conf.* VSP, Utrecht, 1997, pp. 31–34.
18. Drmota M., Gittenberger B. On the profile of random trees. *Random Structures and Algorithms*, **10**, 1997, 421–451.

19. Duran B. S., Odell P. L. *Cluster Analysis. A survey.* Springer–Verlag, Berlin – Heidelberg – New York, 1974.
20. Egorova I. A. The distribution of vertices in strata of plane planted forest. *Probabilistic Methods in Discrete Mathematics. Proc. IV Intern. Petrozavodsk Conf.* VSP, Utrecht, 1997, pp. 179–188.
21. Erdös P., Spenser J. *Probabilistic Methods in Combinatorics.* Academic Press, New York, 1974.
22. Gnedenko B. V., Kolmogorov A. N. *Limit distributions for sums of independent random variables.* Addison–Wesley, Reading, MA, 1954.
23. Flajolet P., Odlyzko A. M. Random mapping statistics. Advances in Cryptology: Proc. of Eurocript'89. Springer. Lecture Notes in Computer Science, **434**, 1990, 329–354.
24. Goncharov V. L. On the distribution of cycles in permutations. *Soviet Math. Dokl.*, **35**, No. 7, 1942, 299–301 (in Russian).
25. Goncharov V. L. On the field of combinatorics. *Soviet Math. Isv.*, **8**, 1944, 3–48 (in Russian).
26. Harrary F. *Graph Theory.* Addison–Wesley, London, 1969.
27. Harrary F. Palmer E. M. *Graphical Enumeration.* Academic Press. New York and London, 1973.
28. Ibragimov I. A., Linnik Yu. V. *Independent and stationary sequences of random variables.* Wolters–Noordhoof, Groningen, 1971.
29. Kalinina N. B., Pavlov Yu. L. The distribution of the degrees of vertices of a random forest. *Branching processes.* Karelian Branch Soviet Acad. Sci., Petrozavodsk, 1981, pp. 10–16 (in Russian).
30. Kalugin I. B. Branching processes and random mappings of finite sets. *Math. Notes*, **34**, 1983, 757–771 (in Russian).
31. Kalugin I. B. The height of random mapping. *Random processes and its applications.* MIEM, Moscow, 1983, pp. 136–144 (in Russian).
32. Kalugin I. B. A class of random mappings. *Proc. of the Steklov Institute of Mathematics*, **177**, No. 4, 1988, 79–110.
33. Kennedy D. P. The Galton–Watson process conditioned on the total progeny. J. Appl. Probab., **12**, No. 4, 1975, 800–806.
34. Kersting G. *Symmetry properties of binary branching trees.* Preprint. Fachbereich Mathematik. Univ. Frankfurt, 1997.
35. Kesten H., Pittel B. Local limit theorems for the number of nodes, the height and the number of leaves in a critical branching process tree. *Random Structures and Algorithms*, **8**, 1996, 243–299.
36. Kolchin V. F. Branching processes, random trees and generalized scheme of arrangemets of particles. *Math. Notes*, **21**, 1977, 386–394.
37. Kolchin V. F. The moment of extinction of a branching process and the height of a random tree. *Math. Notes*, **24**, 1978, 954–961.
38. Kolchin V. F. *Random Mappings.* Springer, New York, 1986.
39. Kolchin V. F. On the behaviour of a random graph near a critical point. *Theory Probab. Appl.* **31**, 1986, 439–451.
40. Kolchin V. F. On Goncharov's work in the field of combinatorics. *Probabilistic methods in discrete mathematics. Proc. III Intern. Petrozavodsk Conf.* VSP/TVP, Utrecht/Moscow, 1993, pp. 23–27.

41. Kolchin V. F. Random graphs and systems of linear equations in finite fields. *Random Structures and Algorithms*, **5**, 1994, 135–146.
42. Kolchin V. F. Systems of random linear equations with small number of nonzero coefficients in finite fields. *Probabilistic methods in discrete mathematics. Proc. IV Intern. Petrozavodsk Conf.* VSP, Utrecht, 1997, pp. 295–304.
43. Kolchin V. F., Sevastyanov B. A., Chistyakov V. P. *Random Allocations.* Wiley, New York, 1978.
44. Kovalenko I. N., Levitskaya A. A., Savchuk M. N. *Selected Problems of the Probability Combinatorics.* Naukova Dumka, Kiev, 1986 (in Russian).
45. Le Gall J.-F. Branching processes, random trees and interacting particle systems. *7th Vilnius Conf. on Proab. Theory. 22nd European Meeting of Statisticians. Abstracts.* TEV, Vilnius, 1998, p. 9.
46. Lyons R., Peres Y. *Probability on trees and networks.* Book in preparation, available at `http://php.indiana.edu/~rdlyons/prbtree/ prbtree.html`
47. Luczak T., Pittel B. Components of random forests. *Combinatorics Probability and Computing*, **1**, 1992, 35–52.
48. Mahmoud H. M. *Evolution of random search trees.* Wiley, New York, 1992.
49. Meir A., Moon J. W. On the altitude of nodes in random trees. *Can. J. Math.*, **30**, No. 5, 1978, 997–1015.
50. Muhin A. B. Local limit theorems for lattice random variables. *Theory Probab. Appl.*, **36**, 1991, 660–674 (in Russian).
51. Palmer E. M. *Graphical Evolution: an introduction to the theory of random graphs.* Wiley, New York, 1985.
52. Pavlov Yu. L. Limit distributions of the number of trees of a given size in a random forest. *Soviet Math. Sb.*, **32**, 1977, 335–345.
53. Pavlov Yu. L. The asymptotic distribution of the maximum tree size in a random forest. *Theory Probab. Appl.*, **22**, 1977, 509–520.
54. Pavlov Yu. L. A case of limit distribution of the maximum size of a tree in a random forest. *Math. Notes*, **25**, 1979, 387–392.
55. Pavlov Yu. L. Random forests and one of branching processes problem. *Mathematical problems of modelling of complicated objects.* Karelian Branch Soviet Acad. Sci., Petrozavodsk, 1979, 41–48 (in Russian).
56. Pavlov Yu. L. The local limit theorem with variable lattice for one characteristic of a random forest. *Branching processes.* Karelian Branch Soviet Acad. Sci., 1981, 24–29.
57. Pavlov Yu. L. Limit distributions of the height of a random forest. *Theory Probab. Appl.*, **28**, 1983, 471–480.
58. Pavlov Yu. L. On the distribution of the number of vertices in strata of a random forest. *Proc. 1st World Congress Bernulli Society*, Vol. 1, VNU Science Press, Utrecht, 1987, 239–241.
59. Pavlov Yu. L. On random mappings with constraints on the number of cycles. *Proc. of the Steklov Institute of Mathematics*, **177**, No. 4, 1988, 131–142.
60. Pavlov Yu. L. On the distributions of the number of vertices in strata of a random forest. *Theory Probab. Appl.*, **33**, No. 1, 1988, 105–114 (in Russian).
61. Pavlov Yu. L. Some properties of plane planted trees. *Discrete Math. and its Applications for Modelling of Complicated Systems. Abstracts.* Irkutsk State Univ., Irkutsk, 1991, 14 (in Russian).

62. Pavlov Yu. L. Some properties of planar planted trees. *Discrete Math. Appl.*, **3**, 1993, 97–102.
63. Pavlov Yu. L. Limit distributions of the height of a random forests consisting of plane rooted trees. *Discrete Math. Appl.*, **4**, 1994, 73–88.
64. Pavlov Yu. L. Asymptotic behaviour of the random forest height. *Proc. of the Department of Mathematics and Data Analysis of Karelian Research Centre of the Russian Acad. Sci.*, **1**, Petrozavodsk, 1994, 4–17 (in Russian).
65. Pavlov Yu. L. The limit distribution of the maximum size of a tree in a random forest. *Discrete Math. Appl.*, **5**, No. 4, 1995, 301–316.
66. Pavlov Yu. L. *Random forests.* Karelian Research Centre of the Russian Acad. Sci., Petrozavodsk, 1996 (in Russian).
67. Pavlov Yu. L. Limit distributions of the number of trees of a given size in a random forest. *Discrete Math. Appl.*, **6**, No. 2, 1996, 117–134.
68. Pavlov Yu. L. Random forests. *Probabilistic Methods in Discrete Mathematics. Proc.* IV *Intern. Petrozavodsk Conf.*, VSP, Utrecht, 1997, pp. 11–18.
69. Pavlov Yu. L., Cheplyukova I. A. Limit distributions of the number of vertices in the layers of a simply generated forest. *Discrete Math. Appl.*, **11**, No. 1, 1999, 97–112 (in Russian).
70. Pavlov Yu. L., Sergeeva I. Yu. The convergence to the limit distributions of the one characteristic of a random forest. *Probabilistic problems of applied mathematics.* Petrozavodsk State Univ., Petrozavodsk, 1984, 48–53 (in Russian).
71. Pitman J. Enumerations of trees and forests related to branching processes and random walks. In. D. Aldous and J. Propp, editors, *Microsurveys in Discrete Probability*, No. 41 in DIMACS Ser. Discrete Math. Theoret. Comp. Sci. Providence RI, Amer. Math. Soc., 1998, 163–180.
72. Pitman J. Coalescent random forests. *J. Comb. Theory*, ser. A, **85**, No. 2, 1999, 165–193.
73. Polya G. Kombinatoriche Anzahlbestimmungen für Grüppen, Graphen und chemische Verbindungen. *Acta Math.*, **68**, 1937, 145–254.
74. *Probabilistic Methods in Discrete Mathematics. Proc. III Intern. Petrozavodsk Conf.* Ed. V. F. Kolchin, V. Ya. Kozlov, Yu. L. Pavlov, Yu. V. Prokhorov. VSP/TVP, Utrecht/Moscow, 1993.
75. *Probabilistic Methods in Discrete Mathematics. Proc. IV Intern. Petrozavodsk Conf.* Ed. V. F. Kolchin, V. Ya. Kozlov, Yu. L. Pavlov, Yu. V. Prokhorov. VSP, Utrecht, 1997.
76. Prüfer A. Neuer Beweis eines Satzes über Permutationen. *Archiv der Mathem. und Phisik*, **27**, 1918, 742–744.
77. *Reference book of probability theory and mathematical statistics.* Nauka, Moscow, 1985 (in Russian).
78. Sachkov V. N. *Probabilistic Methods in Combinatorial Analysis*, Nauka, Moscow, 1978 (in Russian).
79. Sachkov V. N. *Introduction to Combinatorial Methods of Discrete Mathematics.* Nauka, Moscow, 1982 (in Russian).
80. Sachkov V. N. *Combinatorial Methods in Discrete Mathematics.* Cambridge Univ. Press, Cambridge, 1996.
81. Sevastyanov B. A. *Branching Processes.* Nauka, Moscow, 1971 (in Russian).

82. Shapiro A. A generalized distribution model for random recursive trees. *Acta Informatica*, **34**, 1997, 211–216.
83. Stepanov V. E. On the distribution of the number of vertices in strata of a random tree. *Theory Probab. Appl.*, **14**, No. 1, 1969, 65–78.
84. Stepanov V. E. The limit distributions of certain characteristics of random mappings. *Theory Probab. Appl.*, **14**, No. 4, 1969, 612–626.
85. Stepanov V. E. On some features of the structure of a random graph near a critical point. *Theory Probab. Appl.*, **32**, 1988, 573–594.
86. Tkachuk S. G. Local limit theorems for large deviations in the case of stable limiting laws. *Isv. Akad. Nauk* UzSSR, Ser. fiz.-mat. nauk, **2**, 1973, 30–33 (in Russian).
87. Vatutin V. A. The distribution of the distance to the root of the minimal subtree consisting all vertices of a given height. *Theory Probab. Appl.*, **38**, 1993, 273–287 (in Russian).
88. Viskov O. V. Some remarks about branching processes. *Math. Notes*, **8**, No. 4, 1970, 409–418.
89. Zemlyachenko V. N., Pavlov Yu. L. Forests of plane planted trees and branching processes. *Appl. Math. Informatics*, **1**, Petrozavodsk State Univ., Petrozavodsk, 1992, 130–135 (in Russian).

Bei Fragen zur Produktsicherheit wenden Sie sich bitte an:
If you have any questions regarding product safety,
please contact:

Walter de Gruyter GmbH
Genthiner Straße 13
10785 Berlin
productsafety@degruyterbrill.com